Exploring hyperspace
A non-mathematical explanation of multivariate analysis

Exploring hyperspace
A non-mathematical explanation of multivariate analysis

Mick Alt

McGRAW-HILL BOOK COMPANY

London · New York · St Louis · San Francisco · Auckland · Bogotá · Guatemala · Hamburg · Lisbon · Madrid · Mexico · Montreal · New Delhi · Panama · Paris · San Juan · São Paulo · Singapore · Sydney · Tokyo · Toronto

Published by
McGRAW-HILL Book Company (UK) Limited
Shoppenhangers Road · Maidenhead · Berkshire · England SL6 2QL
Tel: 0628-23432
Fax: 0628-35895

British Library Cataloguing in Publication Data
Alt, Mick
  Exploring hyperspace: a non-mathematical explanation of
  multivariate analysis.
  1. Multivariate analysis
  I. Title
  519.5'35

  ISBN 0-07-707338-X

Library of Congress Cataloging-in-Publication Data
Alt, Mick
  Exploring hyperspace: a non-mathematical explanation of
  multivariate analysis/Mick Alt.
      p. cm.
  Includes bibliographical references.
  ISBN 0-07-707338-X
1. Multivariate analysis.   I. Title.
QA278.A46   1990
519.5'35—dc20   89-13686

1234 B&S 9210

Typeset by Rowland Phototypesetting Limited, Bury St Edmunds, Suffolk
and printed and bound in Great Britain by Billing and Sons Limited,
Worcester

For my children
Liam, Niall and Siobhán

Sadly, Mick Alt died during the production of this book, but its publication will ensure that his work and the clarity he brought to complex issues will not disappear.

Contents

1. Some useful statistical ideas   1
2. Introduction to the data set   26
3. Principal components analysis and factor analysis   48
4. Mapping techniques   81
5. Cluster analysis   110
Index   137

Preface

We probably all remember those occasions when we came home from school puzzled by some new aspect of mathematics we'd been 'taught' during the day. If our parents thought they knew about the problems puzzling us, they would produce pen and paper and try to explain to us in a few minutes what, if they could remember, had taken them weeks or months to understand when they were the same age. Often they were surprised when we were unable to understand what they were trying to teach us. If they had thought about the situation a little more deeply, they would have realized that the problem was one of empathy—a failure on *their* part to put *themselves* in *our* shoes and see that ideas, plain as day to them, were unintelligible to us, as beginners.

The 'problem of empathy' is one that seems to dog most experts when they try to teach novices about anything, and in no subject is this more apparent than in mathematics. To accomplished mathematicians, mathematics comes easy, but to the majority it's a hard slog. It is hardly surprising, therefore, that many otherwise intelligent people fight shy of mathematical ideas for the rest of their lives—the scars of bad teaching never heal.

Before outlining what the book is about in a little more detail, a few words about what is *not* included and *why not* might be helpful.

It does not include a bibliography with suggestions for further reading. The reason for this is that to my knowledge other texts that are available on the topic all make the kinds of assumptions about the reader's knowledge that I've tried to avoid. My advice to the reader who wants to take the subject further is first of all to follow a course in matrix algebra. After that, a browse in a university bookshop will reveal a large number of texts from which to choose.

Nor does the book contain a glossary of terms. Whenever a new term is introduced in the text, an attempt has been made to explain it. Personally, I find this more useful than a list of definitions in a glossary. To corrupt a Johnsonian comment, definitions are the last refuge of a scoundrel; what's needed is an *explanation*. Nevertheless, a full index is included to help the reader to cross refer.

There is now a vast number of different software packages available to run on microcomputers that perform the majority of analyses referred to in this book. Recommendation or review of these programmes is beyond the scope of this book (and beyond my own capabilities!)

Nowhere have I mentioned the cost-effectiveness of the different techniques that are explained. If the reader thinks this is an error, then it is most certainly an error of commission. Once a software package has been purchased, the actual cost of carrying out the analyses described is—quite literally—the cost of the electricity to run the computer. In other words, the cost is almost zero. The effectiveness of the techniques depends, in my view, almost entirely on the competence of the researcher carrying out the analyses. Here, I've concentrated on helping researchers to develop a better understanding of the techniques themselves.

Increasingly however, an understanding of ideas developed in mathematical statistics is necessary to practitioners in survey research. In recognition of this demand, the last decade has seen a huge proliferation of books on various statistical topics for the non-statistician. All but the tiniest minority fail to achieve what they set out to—statistics made easy. I'm convinced that the major reason for this state of affairs is, quite simply, that statisticians who write books for non-statisticians are quite unable—no matter how good their intentions—to see that things they take for granted are by no means obvious to the beginner. The problem is that as statisticians get more and more knowledgeable in their subject, ideas that they found difficult to start with become more and more obvious and self-evident. Eventually, as their familiarity with statistical ideas becomes almost 'reflex', the ideas they try to explain are so obvious (to them) that they find it hard to imagine that anybody could have any difficulty whatsoever in understanding them. It's all very well to say that 'this book is written for readers with no more than an elementary knowledge of mathematics', but it is quite a different matter to write a book that fully embraces this sentiment.

When it comes to books about so-called multivariate statistics (which this is, and about which more in a moment), the situation becomes even more hopeless. Not only do statisticians assume a knowledge of *elementary* statistics, but they also feel the need to introduce the beginner to a particular form of algebra—*matrix algebra*—used in the formulation of many multivariate statistical ideas. In the case where a statistician thinks he is being most helpful, he might begin his book with an introductory chapter on matrix algebra. In such a chapter, he might try to condense in a few pages what, for him, formed a substantial part of an advanced course in mathematics. A less helpful statistician 'assumes a rudimentary knowledge of statistics and matrix algebra'. In both instances, these types of books almost invariably fail to communicate the important concepts to non-mathematicians.

In this book, no assumptions are made about the reader's statistical knowledge; no matrix algebra and no mathematical formulae are included. (A fairly *basic* knowledge of secondary school mathematics is assumed.) In this sense, the book is completely self-contained. All statistical terms are introduced and explained as, or before, they are needed, although the first chapter also contains some useful statistical ideas worth taking time to understand.

My audience is most definitely not statisticians, though I hope statisticians are not too upset with any of the liberties they might feel I've taken with their subject. My aim has been to simplify without trivializing. The audience for whom I've written comprise survey research executives in the public and private sectors, research buyers, marketing executives who want to know something about the techniques used to derive the information upon which they make decisions, and students in business studies and the social sciences.

I mentioned a few paragraphs back that this is a book about methods of multivariate analysis. What is meant by multivariate analysis? Most of us are familiar with *univariate* statistics, the examination of a single variable being studied, such as the distribution of salaries in a company. Many of us are also familiar with *bivariate* statistics—the examination of two variables simultaneously, such as the relationship between salaries and age of employees in a company. Multivariate statistics is concerned with the examination of three or more variables simultaneously, for example the relationship between salaries, age, gender, educational qualifications and ethnic background of employees in a company. Most of the data in survey research are, by necessity, multivariate since the technique usually involves measuring or counting many characteristics about members of a population. A preliminary way of examining multivariate data is to take the variables in pairs, covering all possible pairs, and examining the relationships between the pairs in the form of a graph or *scatterplot* (see Chapter 1). This is not entirely satisfactory. To examine the relationship between three variables *at the same time* we could draw a 'three-dimensional' picture, but with more than three variables, we are stymied. It requires what is known as a *multidimensional space* or *hyperspace* to examine more than three variables simultaneously. These are mathematical and not visual concepts. This book is about hyperspace and some of the methods used to explore it. It is not a comprehensive coverage and the aim is to give the reader a comparatively easy way of understanding how some of the more popular methods of analysis work. More specifically, the book covers three main methods of multivariate analysis, namely, factor analysis, mapping methods and cluster analysis.

The first chapter is concerned with introducing some very useful statistical ideas. These ideas form the building blocks for the later chapters but are well worth knowing about in their own right. The second chapter introduces a particular study and the data arising from it; these form the basis from which the methods of analysis covered in the remaining chapters are discussed. In addition, some useful and easy ways of examining the data without recourse to complicated mathematics are illustrated. Chapter 3 deals with factor analysis; Chapter 4 with mapping methods and the final chapter with cluster analysis. None of the chapters is completely self-contained. Many of the ideas introduced in earlier chapters are followed up and used in subsequent chapters. Finally the book is written as a narrative—Chapter 1 is very clearly the beginning, and each of the following chapters takes the story further.

Mick Alt

Acknowledgements

My thanks are due to Peter Finer of Finer, Weston Associates for helping me with some of the preliminary analysis using his remarkable survey analysis package that runs on a Mackintosh microcomputer. I also used his program for the principal components analysis reported in Chapter 3, for which I am grateful.

Thanks are also due to David Dawkins of Probit Research Ltd for running the correspondence analysis referred to in Chapter 4 and for his comments on this chapter. Likewise, I'd like to thank Paul Harris of NOP Ltd for running the 'iterative partitioning' cluster analysis described in Chapter 5 and also for his comments on the chapter.

Eric Willson of Digitab Ltd kindly read the entire manuscript and I am grateful for his comments and suggestions.

My son, Liam, suggested the design for the book cover and I'd like to thank him since it proves that a ten year old can grasp the idea of hyperspace! Finally, I'd like to thank Mini Tandon and Bassy Okon of Creative Research Ltd for their care in preparing the typescript.

Mick Alt
Creative Research Ltd
London

1. Some useful statistical ideas

Statistics usually starts with a set of measurements on the members of a population. As an example, Table 1.1 shows the weather conditions in 23 towns designated as belonging to the south coast of England. Inspection of the table shows that at the time of writing this chapter in a rather damp and cold London, the *range* of temperatures in south coast towns was between 52 and 62 °F.

A better way of grasping the total picture regarding these temperatures is to arrange the figures from the 'Temperature' column of Table 1.1 in the form of a *histogram*. This is shown in Fig. 1.1, in which the frequency of the average daily temperatures is represented by the height of a column with the temperature at the centre of the base. This has been done using degrees Fahrenheit rather than degrees Celsius.

If you examine the frequency columns in this histogram, you'll see that the most common temperature on the English south coast during the autumn day in question was 59 °F and that the frequencies of temperatures above and below 59 °F tend to diminish as you move either side of 59 °F.

The arithmetic average temperature of south coast towns, or the *mean*, as statisticians call it to distinguish it from averages defined differently, is calculated simply enough by adding together the 23 temperatures and dividing this sum by 23. This comes to 58.96 °F, which is close enough to the most frequently occurring temperature, shown in Fig. 1.1.

So far, the data in Table 1.1 have been described by two statistics

Table 1.1
Weather in English towns on an autumn day in 1988

South coast town	Hours of sunshine	Inches of rain	Temperature (°C)	(°F)	Weather (day)
Folkestone	—	—	13.5	56	Cloudy
Hastings	0.1	—	14	57	Cloudy
Eastbourne	2.4	—	14.5	58	Bright
Brighton	0.1	—	14	57	Cloudy
Worthing	1.2	—	14.5	58	Cloudy
Littlehampton	1.4	—	14	57	Cloudy
Bognor Regis	1.4	—	14.5	58	Bright
Southsea	1.9	—	15	59	Cloudy
Sandown	1.8	—	15	59	Bright
Shanklin	0.8	—	15	59	Cloudy
Ventnor	2.3	—	15	59	Bright
Bournemouth	3.8	—	15.5	60	Sunny p.m.
Poole	4.4	—	16	61	Sunny p.m.
Swanage	3.8	—	15	59	Sunny p.m.
Weymouth	3.9	0.01	15	59	Sunny p.m.
Exmouth	3.1	0.04	15	59	Sunny a.m.
Teignmouth	5.0	0.14	15.5	60	Sunny a.m.
Torquay	3.7	0.15	15.5	60	Bright a.m.
Falmouth	3.7	0.23	15.5	60	Showers
Penzance	4.1	0.24	15.5	60	Showers
Isles of Scilly	2.4	0.26	14.5	58	Sunny p.m.
Jersey	5.0	0.04	16	61	Bright
Guernsey	5.8	—	16.5	62	Sunny a.m.

—the temperature range, 56–62 °F, and the mean temperature, 58.96 °F. Are these two statistics sufficient to summarize the data? The answer, unfortunately, is no; this is because neither statistic tells us very much about the *scatter* or *dispersion* around the mean. For example, instead of the actual data represented in Fig. 1.1, imagine there were only three observed temperatures in the 23 south coast towns: seven of 59 °F (as, indeed, in Table 1.1), eight of 56 °F and eight of 62 °F. This would give us a range of 56–62 °F, the same as before, and a mean temperature of exactly 59 °F. Although the ranges and means in these two cases are nearly identical, the scatter or dispersion of scores around the means is profoundly different. To complete the picture, then, some measure of the scatter of the scores is needed. The statisticians' jargon for such a measure is to speak of measurements *deviating* from the mean. One might simply calculate how much the individual towns' temperatures differ from the mean, and then take the mean of these differences, once they've all been added together. Unfortunately, this simple solution has awkward arithmetical consequences, not the least of which is the fact that some of the deviations are negative and some are positive, such that when they are added together, they amount to nought. In fact, upon reflection, this should not be too surprising, for if the mean is to be numerically representative, then the sum of the positive and negative deviations of the individual values round it must—by definition—equate to produce a zero answer.

To get round this awkward arithmetical fact, it is usual for statisticians to talk about the 'mean of the squared deviations from the mean', which is known as *variance*. This is an index of *variability* much used by statisticians. Let's try and bring this rather abstract discussion to life. Table 1.2 shows, in its first column, the temperatures of south coast towns; in its second column it shows the deviations from the mean (which for ease of calculation is taken to be exactly 59 °F) and in its third column it shows the 'squared deviations' from the mean. The final row in the table shows the mean of the sum of each of these three columns.

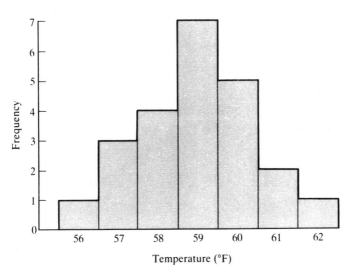

Figure 1.1 *Histogram showing the temperatures in English south coast towns on an autumn day in 1988.*

Table 1.2
Deviations from the mean temperature in south coast towns

South coast town	Temperature (°F)	Deviations from the mean	Squared deviations from the mean
Folkestone	56	−3	9
Hastings	57	−2	4
Eastbourne	58	−1	1
Brighton	57	−2	4
Worthing	58	−1	1
Littlehampton	57	−2	4
Bognor Regis	58	−1	1
Southsea	59	0	0
Sandown	59	0	0
Shanklin	59	0	0
Ventnor	59	0	0
Bournemouth	60	+1	1
Poole	61	+2	4
Swanage	59	0	0
Weymouth	59	0	0
Exmouth	59	0	0
Teignmouth	60	+1	1
Torquay	60	+1	1
Falmouth	60	+1	1
Penzance	60	+1	1
Isles of Scilly	58	−1	1
Jersey	61	+2	4
Guernsey	62	+3	9
TOTAL =	1356	−1	47
MEAN $\simeq$*	59	0	2

*$\simeq$ is a mathematical symbol meaning 'approximately equal to'.

Although the 'mean of squared deviations from the mean' has useful arithmetical properties, it gives rather a false impression at first sight, because we are not used to having to deal with an ordinary number, the mean, and a squared number, the index of the scatter, at one and the same time. It has therefore become a convention to quote the index of scatter as the square root of the 'mean of the squared deviations'. In the jargon, this is known as the *standard deviation*, and it gives an immediately graspable idea of the variability or scatter of the measurements. The mean temperature is 59 °F; the variability, expressed as variance, or mean squared deviations, is two 'square degrees Fahrenheit', which is not very easy to grasp. But the same variability, expressed as 'standard deviation', would be the square root of two 'square degrees Fahrenheit', that is, 1.4 °F. This is much easier to understand. To get a more concrete idea from it, all we need to know is what the standard deviation corresponds to in terms of numbers of measurements. This has been worked out in detail by statisticians, who have calculated the properties of a theoretical distribution, known as the *normal distribution*. Examples of four normal distributions are given in Fig. 1.2. These are rather like the histogram of Fig. 1.1 in their general shape, but 'smoothed out' to get rid of the jagged edges. However, each shows a different pattern of scatter about the same mean. The major visual characteristics of normal curves are unimodality (one peak) and symmetry (one side the same as the other). They also have certain mathematical properties such that if the mean and standard deviations of a normal distribution are known, it is possible to draw with deadly accuracy the shape of the curve.

Some of the important mathematical properties of the normal distribution are shown in Fig. 1.3. Here we can see that approximately 68 per cent of all values of a normally distributed variable lie between ±1 standard deviation from the mean; approximately 95 per cent of all values lie between ±2 standard deviations from the mean and over 99 per cent lie between ±3 standard deviations from the mean. According to statistical theory, therefore, if the histogram in Fig. 1.1 is approximately normal, we would expect approx-

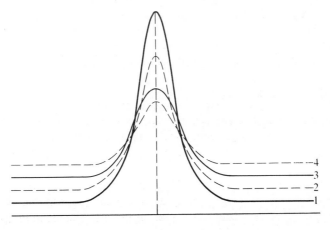

Figure 1.2   *Four normal distributions with the same means but different scatters.*

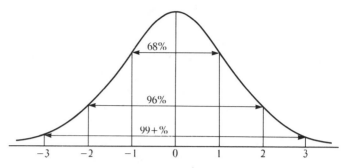

Figure 1.3 *Some important mathematical properties of the normal distribution.*

imately 68 per cent of all temperatures to lie between ±1 standard deviation from the mean temperature, approximately 95 per cent of all temperatures to lie between ±2 standard deviations from the mean temperature and approximately 99 per cent of all temperatures to lie between ±3 standard deviations from the mean temperature. If you remember, the calculated mean and standard deviation are approximately 59 °F and 1.4 °F, respectively. Therefore, if the distribution of temperatures in south coast towns is approximately normal, we would expect approximately 68 per cent of all temperatures to lie in the range 59 °F ± 1.4 °F, that is, between 57.6 °F and 60.4 °F. Rounding these figures to whole numbers, the expected range would be 58–60 °F. In fact, an inspection of Fig. 1.1 shows that $^{16}/_{23}$, or 70 per cent of all temperatures lie within the range. Not bad! Applying comparable calculations based on the theoretical assumptions of the normal distribution just discussed, we can show, mathematically, what we can see visually, namely, that Fig. 1.1 approximates to a normal distribution. So what? For survey researchers, one of the most important consequences of the properties of the normal distribution relates to questions about *samples* drawn from a population. We'll look at this in a little more detail in the next section. Before leaving this section, however, it is also worth noting that any variable which is normally distributed can be converted into what are known as *standard scores*, or *z* scores.

For example, in Table 1.1, the temperatures of south coast towns are shown in degrees Fahrenheit and also in degrees Celsius. Forget for a moment that it is a relatively simple matter to convert Fahrenheit to Celsius and vice versa, and imagine you wanted to compare the temperatures of the towns on two separate occasions. On one occasion the temperatures were measured in degrees

Fahrenheit and on the other in degrees Celsius, but you didn't know which was which. Moreover, imagine that all you know is that the temperatures on these two occasions are measured on two different unknown scales in units of measurement that can be converted precisely, one to the other.

By converting the measurements to standard scores, it is possible to make direct comparisons, irrespective of the actual units of measurements used. Standard scores are those in which the mean of the population is equal to zero and the standard deviation is equal to unity; and therefore they range from -3 standard deviations through 0 to $+3$ standard deviations.

Any normally distributed variable can be transformed to its standardized equivalent. To convert each temperature in Table 1.1 measured in degrees Fahrenheit to standard scores, you subtract the mean and divide by the standard deviation. So, for example, a temperature of 56 °F, expressed as a standard score is $(56 - 59)/1.4$, which is approximately equal to -2; a temperature of 62 °F is approximately equal to $+2$ and the mean temperature is equal to 0. If you are inclined to carry out the necessary computations, you can convince yourself that the corresponding scores, measured in degrees Celsius, and the standardized scores are the same. First of all, you'll have to compute the mean temperature and standard deviation in degrees Celsius!

Of course, not all variables in a population are normally distributed, as temperature is in the population of south coast towns. Indeed, in this population, the variable 'inches of rain' is far from normally distributed, and statistically this can cause considerable problems when one tries to summarize the variables in question. It can become a very technical job which we shall not pursue here. But it is worth pointing out that unless a population variable is normally distributed, the mean can give a wrong impression if it is taken as a measure of the average. An example will make this important point crystal clear. Figure 1.4 shows the distribution of salaries in a company. Here you can see that the mean salary gives a very misleading indication of the distribution of salaries in the company. A few employees on relatively high salaries tend to pull the arithmetic average up. In this example, much better indications of the average salary are given by the *median* and the *mode* (defined in Fig. 1.4).

Unless you know that a variable is normally distributed, it often pays to calculate the median value and the mode as well as the

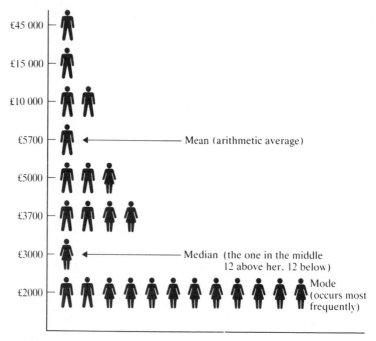

Figure 1.4 *Distribution of salaries in a company.*

mean. The mean is often used inappropriately, partly because it allows the researcher to carry out lots of statistical analyses which are just not possible on other measures of the average. (No doubt, you will have noticed that in a normal distribution the mean = the median = the mode.)

Samples

So far, it has been assumed implicitly that one can measure every member of the population. In practice, in survey research, this is almost never the case. Typically, what we do is take a *sample* from the population. A good many of the most important practical ideas of statistics are concerned with sampling. The most basic thing to realize is that a sample cannot provide a *precise* picture about a population. So, one of the things researchers do is to take great care to ensure that any sample they take is *unbiased*—a *random* sample, as it is usually called. (We'll take a look at what is meant by random in the next section.) In practical terms, the questions we are likely to be interested in are either how accurate a picture can we expect from a given sample, or how large does the sample have to be to provide us with a specified degree of accuracy? The answer to both these questions is provided by a statistic known as the *standard error* of the mean. The best way to get a feeling for the standard error is to carry out what Einstein called a 'thought experiment'. (Incidentally, these are the only experiments Einstein ever carried out.) Here goes.

Imagine you selected a random sample of 400 housewives and you asked each of them whether they had ever used a particular product. Suppose 120 housewives in your sample said 'yes' and 280 said 'no'. By simple arithmetic, you'd estimate, on average, ($^{120}/_{400} \times 100$) per cent, i.e. 30 per cent, of housewives had ever used the product in question. Our interest here is how reliable is this average value of 30 per cent as an estimate of usership among the population at large? Imagine you selected another random sample of 400 housewives and you asked the same question. You would not expect to get an answer of exactly 30 per cent this time, but you'd hope to get a figure something near it. Imagine you repeated this exercise over and over again and you plotted the results (i.e. the percentage of housewives ever using the product) in the form of a histogram like the one shown in Fig. 1.1. What would it show you? Whenever this has been done in comparable situations (e.g. repeatedly tossing a coin 400 times), it has been found that the histogram tends to be symmetric around the mean of all the sample means. The standard deviation of this distribution of sample means is known as the standard error of the mean. Luckily there is no need to draw a large number of different random samples to estimate the standard error.

In our example, the standard error can be estimated as follows:

$$\sqrt{\left(\frac{30\% \times 70\%}{400} \right)} = 2.29\%$$

In the general case, when dealing with percentages, to calculate the standard error, you multiply the percentage owning the product (or, more generally, possessing some characteristic) by the percentage not possessing it (this product in fact gives you an estimate of the variance); you divide this product by the sample size, and the square root of the resultant figure provides an estimate of the standard error.

Now, the important thing to grasp is that the standard error is a standard deviation and the distribution of the sample means is normal. (This latter point is true even when the variable being measured is not normally distributed, as is the case in our example of product usership.) These two properties permit us to interpret the standard error exactly as we would any standard deviation of a normal distribution. It was pointed out in Fig. 1.3 that 95 per cent of all the values in a normal distribution lie between ±2 standard deviations from the mean. Let's now apply this finding to our example of the survey of 400 housewives, 30 per cent of whom had ever used the product in question, with an estimated standard error of 2.29 per cent. We can say we would expect the proportions using the product to lie between 30± (2 × 2.29) per cent, that is, between 25.42 and 34.58 per cent, 95 per cent of the time, in the long run.

The limits of bounds of the 'true' percentages calculated in this way are called *confidence limits*. The width of the confidence limits indicates the accuracy of the percentages derived from the sample. Obviously, the closer the limits, the more accurate the estimate is. One way of reducing the width is to reduce the size of the standard error, and one obvious way of reducing the standard error is to increase the sample size. Indeed, as the sample size is increased, the standard error tends towards zero until it reaches zero! This is when everybody in the population is sampled, and the result is called a *census*.

Table 1.3 shows the 95 per cent confidence limits associated with different sample sizes and different percentages. Two obvious trends can be seen. Firstly, by looking along any of the rows, the width of the confidence limits increases as the percentage approaches 50 per cent and decreases the further from 50 per cent it moves. Secondly, by looking down any of the columns, the width of the confidence limits decreases as the sample sizes increase. So, an answer to the question 'How big a sample do I need?' depends on two factors, namely, the percentage of people giving an answer one way or the other and the sample size. Usually, one doesn't have any information about the former—this is probably the reason why one carries out a survey in the first place—so it's usual to take the worst possible case (the final column in Table 1.3). Notice, however, that as the sample size increases a law of diminishing returns sets in, such that a point is reached where large increases in sample size result in small changes in the width of the confidence limits, and hence in the precision of the quantity being estimated. In deciding on the sample

Table 1.3
Table of confidence limits

Sample size	Percentage giving a particular answer/possessing a particular characteristic									
	5% 95%	10% 90%	15% 85%	20% 80%	25% 75%	30% 70%	35% 65%	40% 60%	45% 55%	50% 50%
100	4.4	6.0	7.1	8.0	8.7	9.2	9.5	9.8	9.9	10.0
150	3.6	4.9	5.8	6.5	7.1	7.5	7.8	8.0	8.1	8.2
200	3.1	4.2	5.0	5.7	6.1	6.5	6.7	6.9	7.0	7.1
250	2.8	3.8	4.5	5.1	5.5	5.8	6.0	6.2	6.3	6.3
300	2.5	3.5	4.1	4.6	5.0	5.3	5.5	5.7	5.7	5.8
400	2.2	3.0	3.6	4.0	4.3	4.6	4.8	4.9	5.0	5.0
500	1.9	2.7	3.2	3.6	3.9	4.1	4.3	4.4	4.4	4.5
600	1.8	2.4	2.9	3.3	3.5	3.7	3.9	4.0	4.1	4.1
700	1.6	2.3	2.7	3.0	3.3	3.5	3.6	3.7	3.8	3.8
800	1.6	2.1	2.5	2.8	3.1	3.2	3.4	3.5	3.5	3.5
900	1.5	2.0	2.4	2.7	2.9	3.1	3.2	3.3	3.3	3.3
1000	1.4	1.9	2.3	2.5	2.7	2.9	3.0	3.1	3.1	3.2
1500	1.1	1.5	1.8	2.1	2.2	2.4	2.5	2.5	2.6	2.6
2000	1.0	1.3	1.6	1.8	1.9	2.0	2.1	2.2	2.2	2.2
3000	0.8	1.1	1.3	1.5	1.6	1.7	1.7	1.8	1.8	1.8
4000	0.7	0.9	1.1	1.3	1.4	1.4	1.5	1.5	1.6	1.6
5000	0.6	0.8	1.0	1.1	1.2	1.3	1.3	1.4	1.4	1.4

size, you must decide on the precision you require. This is a question that cannot be answered by statistics and it depends on the purpose to which the results of the survey will be put. Statistical theory can only tell you how precise your answer is, not how precise it should be.

Up to now, we have looked at only one side of the typical decision regarding sample size, that is, what size of sample is required to yield a specified precision in estimating some quantity. In practice, the aim is either to get maximum precision at a certain cost (or in the time available) or to reach a predetermined level of precision at the lowest cost. So, a decision on sample size is as much governed by an estimate of the costs and time involved as it is by the confidence limits.

In many surveys, the researcher is not only concerned with obtaining an estimate of some quantity within acceptable limits of precision but also with comparing differences between the average values of different groups as measured on some variables.

Suppose in our example of the 400 housewives we were interested in finding out whether younger housewives were more likely to have ever used the product than older housewives. To find this out, we might 'breakdown' the total percentage who had ever used the product (30 per cent) into the percentages of younger and older housewives. The table produced by this process might look something like Table 1.4. From this table, we can see that the difference in usage of the product between younger and older housewives is 16 per cent. The question we ought to ask ourselves is whether this difference of 16 per cent is a *real* difference or merely due to sampling error. An answer to this question is provided by the confidence limits associated with the answers of the younger and older housewives. There is a significance test we could apply to the data in Table 1.4 but for most purposes a rule of thumb will do. Firstly, reference to Table 1.3 will show us the confidence limits associated with the answers of younger and older housewives. The confidence limit associated with the younger housewives is 8 per cent; so in other words the 'true' percentages of younger housewives ever having used the product lies in the range 32–48 per cent. The confidence limit associated with the older housewives is approximately 5.8 per cent; so, in other words the 'true' percentage of older housewives who have ever used the product lies in the range 18.2–29.8 per cent. As these two ranges do not overlap, we may

Table 1.4

Percentages of housewives ever using a particular product

	Total (400) %	Younger housewives (150) %	Older housewives (250) %
Ever used	30	40	24
Never used	70	60	76

conclude with reasonable confidence that younger housewives are more likely to have ever used the product than older housewives. This general procedure can be translated into the following rule of thumb.

If the larger margin of error (confidence limit) is less than the observed difference, then the observed difference is statistically significant at the 95 per cent confidence level; if the larger margin of error is greater than the observed difference, the observed difference is not statistically significant at the 95 per cent confidence level.

In concluding this section, it is worth remembering that the only general principle about assessing the reliability of measurements or estimates is that they can never be 100 per cent reliable. All that the most elaborate statistics can do is give you something like betting odds for or against the hunch that something is, or is not, so. It leaves you to decide whether to bet, and how much to stake.

Randomness and random sampling

Mentions of *random* have been made on a few occasions throughout this chapter. So, perhaps it is worth while to consider the ideas of random and randomness in a little more detail.

The notion of randomness is at the core of modern probabilistic methods which are themselves central to all scientific investigations. For example, early this century, randomness became the bedrock of quantum mechanics, and it is at the root of modern evolutionary theory. This crucial scientific idea has also had an influence on the random art of abstract expressionism, in the random music of such composers as John Cage and in the random wordplay of a book such as James Joyce's *Finnegans Wake*. The idea of randomness has had a profound effect on twentieth-century thought, yet it is extremely difficult to state what we mean by 'random'. The dictionary definition of 'haphazard, accidental, without aim or direction' is not very helpful either, for we know that scientists are very systematic about randomness; there is nothing haphazard or accidental in the way they select random samples. Most mathematicians agree that an absolutely disordered series of numbers is a logically contradictory concept. Evidently, the best one can do is to specify certain tests for types of randomness and call a series random to the degree that it passes them. The best way to get a series of random digits (the 'numbers' 0, 1, 2, 3, 4, 5, 6, 7, 8, 9) is to use a physical process involving so many variables that the next digit can never be predicted with a probability higher than $1/n$ where n is the base of the number system being used. Tossing a coin generates a random series of binary digits ($n = 2$). A perfect die randomizes six numerals ($n = 6$), and so on. Nowadays, it is customary to use random numbers generated by a computer rather than tossing a coin or throwing a die each time a series of random numbers is required. Strictly speaking they are known as pseudo-random numbers. This reflects the fact that, philosophically, it is impossible to define randomness.

The idea that a series of numbers is only random when there is no discernible pattern in it is so firmly entrenched in the psyche that it is worth digressing for a moment to squash it. Suppose that somewhere in the very long and apparently patternless string of digits produced by the computer we come across the sequence 0 1 2 3 4 5 6 7 8 9. Our first impression is that at this point the randomizer has broken down, for these ten digits are quite obviously fully patterned and therefore not random. This is, of course, a mistake. The sequence 0 1 2 3 4 5 6 7 8 9 is just as likely as any other combination of ten digits and, indeed, in an 'infinite' series of numbers we would *expect* to see this pattern occur. The sequence appears ordered to us because we have assigned meaning, importance and prominence to this outcome and have bunched all the others together as having no order, as being *random*. We tend to think of randomness as meaning the absence of pattern and therefore whenever we see a pattern we give it special meaning. This concept of randomness bears meaning only in relation to the observer, and if two observers habitually look for different kinds of patterns they are bound to disagree upon the series that they call random.

Remember, we call a series random if we cannot predict successive digits in the series with a probability greater than $1/n$. For example, there is no known way to win a penny-tossing game, in which the probability of any 'toss' being a 'head' or a 'tail' is $1/2$ ($n = 2$). However, we can predict quite well in the aggregate, i.e. we can predict the outcome of large numbers of events. For example, if we toss a coin 10 000 times, we can predict, with considerable accuracy, the total numbers of heads and tails. This apparent contradiction gives rise to what philosophers have called the *gambler's fallacy*. For example, if in tossing a coin you get three 'heads' in a row, the temptation is to predict that on the next toss you will get a 'tail'. This is, of course, faulty reasoning since it is just as likely to be 'heads'.

Having discussed the nature of randomness, we are now in a position to see the importance in sampling. In taking a random sample of the population under study, every member of the population has an equal chance of being selected. A sample drawn at random is *unbiased* in the sense that all possible samples of the same size have an equal chance of being selected. If we do not draw our sample at random, some factor or factors unknown to us may lead us to select a biased sample. Random methods of selection do not allow our own biases or any other systematic selection factors to operate. Random sampling is objective in the sense that it is divorced from an individual's own predilections and biases.

Correlation

A very important part of statistics is concerned with finding out how closely two sets of different things are connected with each other, or, as statisticians say, how *correlated* they are. Again, an example will make this clearer. We can plot a graph showing the relationship between the temperatures and hours of sunshine for each of the south coast towns shown in Table 1.1 by, for example, plotting temperature vertically and hours of sunshine horizontally. The result is known as a *scatterplot*, and the scatterplot produced from the data in Table 1.1 is shown in Fig. 1.5.

There is a clear trend evident from this scatterplot, namely, that on the whole, the hotter it is, the more hours of sunshine there are. From these data it is possible to draw a straight line sloping upwards to the right around which all the points are grouped. There are mathematical procedures for finding the line that fits the points best; and these are known as *regression* analyses. In our example, there is what is known as a *positive* correlation between temperature and hours of sunshine, that is, as one increases so does the other. Sometimes we find variables that are *negatively* correlated: as one variable increases, the other decreases, or vice versa. Negatively correlated variables, are fairly common in biological and ecological systems, for example, in predator–prey relationships. During the summer, you've probably noticed in your own back garden (if you're fortunate enough to have one) that as the population of ladybirds increases, the population of aphids decreases and vice

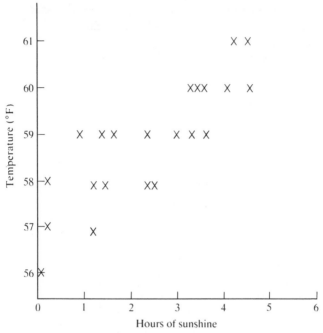

Figure 1.5   *Scatterplot of hours of sunshine and temperature (taken from Table 1.1).*

versa. These two variables (the sizes of the ladybird and aphid populations) are negatively correlated. Correlation coefficients are expressed on a scale from −1 through 0 to +1. A correlation coefficient of −1 would indicate a *perfect* negative correlation, that is, there would be no scatter of points around the line, *all* of them would fall on the line; a correlation coefficient of 0 would indicate that two variables are not at all correlated, and a correlation coefficient of +1 would indicate a perfect positive correlation.

Another way of visualizing what a correlation coefficient signifies can be gained if we imagine a scatterplot produced by variables that have been converted into standard scores. Replotting the scatterplot of Fig. 1.5 in standard scores gives us the scatterplot shown in Fig. 1.6.

Remember transforming scores into standard scores has the effect of creating a zero mean and positive and negative scores (scores above and below the mean, respectively) with a new unit of measurement of one standard deviation. You can see from Fig. 1.6 that the origin of the graph now occurs at zero, the mean for both standard scores; and scores in the first quadrant are all positive while those in the third quadrant are all negative. In other words, as temperature increases, on the whole, so does 'hours of sunshine', and vice versa; and, similarly, as temperature decreases so does 'hours of sunshine'; the converse is also true. When plotted on a scatterplot as standard scores, variables that are negatively correlated will show a scatter of points in the second and fourth quadrants; as one of the variables increases in value, the other variable decreases in value, and vice versa. Lowly correlated variables will show a scatter of points in all quadrants to the extent that when two variables are completely uncorrelated (that is, the correlation coefficient equals zero), an equal number of points will be found in each of the four quadrants.

There is now a large number of different ways of computing the correlation coefficient between two variables and these depend largely on the systems of measurement in which the variables concerned are measured. The correlation coefficient appropriate to two variables, which we've been considering, is known as the *Product Moment Correlation Coefficient*; and when we apply the appropriate formula, which needn't concern us here, we find that

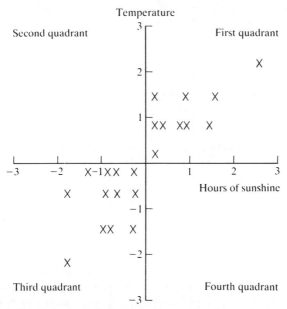

Figure 1.6 *Scatterplot in standard scores of hours of sunshine and temperature.*

Table 1.5
Table of natural cosines

NATURAL COSINES

[Numbers in difference columns to be subtracted, not added.]

Degrees	0' 0°·0	6' 0°·1	12' 0°·2	18' 0°·3	24' 0°·4	30' 0°·5	36' 0°·6	42' 0°·7	48' 0°·8	54' 0°·9	1	2	3	4	5
0	1·000	1·000	1·000	1·000	1·000	1·000	·9999	9999	9999	9999	0	0	0	0	0
1	·9998	9998	9998	9997	9997	9997	9996	9996	9995	9995	0	0	0	0	0
2	·9994	9993	9993	9992	9991	9990	9990	9989	9988	9987	0	0	0	1	1
3	·9986	9985	9984	9983	9982	9981	9980	9979	9978	9977	0	0	1	1	1
4	·9976	9974	9973	9972	9971	9969	9968	9966	9965	9963	0	0	1	1	1
5	·9962	9960	9959	9957	9956	9954	9952	9951	9949	9947	0	1	1	1	2
6	·9945	9943	9942	9940	9938	9936	9934	9932	9930	9928	0	1	1	1	2
7	·9925	9923	9921	9919	9917	9914	9912	9910	9907	9905	0	1	1	2	2
8	·9903	9900	9898	9895	9893	9890	9888	9885	9882	9880	0	1	1	2	2
9	·9877	9874	9871	9869	9866	9863	9860	9857	9854	9851	0	1	1	2	2
10	·9848	9845	9842	9839	9836	9833	9829	9826	9823	9820	1	1	2	2	3
11	·9816	9813	9810	9806	9803	9799	9796	9792	9789	9785	1	1	2	2	3
12	·9781	9778	9774	9770	9767	9763	9759	9755	9751	9748	1	1	2	3	3
13	·9744	9740	9736	9732	9728	9724	9720	9715	9711	9707	1	1	2	3	3
14	·9703	9699	9694	9690	9686	9681	9677	9673	9668	9664	1	1	2	3	4
15	·9659	9655	9650	9646	9641	9636	9632	9627	9622	9617	1	2	2	3	4
16	·9613	9608	9603	9598	9593	9588	9583	9578	9573	9568	1	2	2	3	4
17	·9563	9558	9553	9548	9542	9537	9532	9527	9521	9516	1	2	3	3	4
18	·9511	9505	9500	9494	9489	9483	9478	9472	9466	9461	1	2	3	4	5
19	·9455	9449	9444	9438	9432	9426	9421	9415	9409	9403	1	2	3	4	5
20	·9397	9391	9385	9379	9373	9367	9361	9354	9348	9342	1	2	3	4	5
21	·9336	9330	9323	9317	9311	9304	9298	9291	9285	9278	1	2	3	4	5
22	·9272	9265	9259	9252	9245	9239	9232	9225	9219	9212	1	2	3	4	6
23	·9205	9198	9191	9184	9178	9171	9164	9157	9150	9143	1	2	3	5	6
24	·9135	9128	9121	9114	9107	9100	9092	9085	9078	9070	1	2	4	5	6
25	·9063	9056	9048	9041	9033	9026	9018	9011	9003	8996	1	3	4	5	6
26	·8988	8980	8973	8965	8957	8949	8942	8934	8926	8918	1	3	4	5	6
27	·8910	8902	8894	8886	8878	8870	8862	8854	8846	8838	1	3	4	5	7
28	·8829	8821	8813	8805	8796	8788	8780	8771	8763	8755	1	3	4	6	7
29	·8746	8738	8729	8721	8712	8704	8695	8686	8678	8669	1	3	4	6	7
30	·8660	8652	8643	8634	8625	8616	8607	8599	8590	8581	1	3	4	6	7
31	·8572	8563	8554	8545	8536	8526	8517	8508	8499	8490	2	3	5	6	8
32	·8480	8471	8462	8453	8443	8434	8425	8415	8406	8396	2	3	5	6	8
33	·8387	8377	8368	8358	8348	8339	8329	8320	8310	8300	2	3	5	6	8
34	·8290	8281	8271	8261	8251	8241	8231	8221	8211	8202	2	3	5	7	8
35	·8192	8181	8171	8161	8151	8141	8131	8121	8111	8100	2	3	5	7	8
36	·8090	8080	8070	8059	8049	8039	8028	8018	8007	7997	2	3	5	7	9
37	·7986	7976	7965	7955	7944	7934	7923	7912	7902	7891	2	4	5	7	9
38	·7880	7869	7859	7848	7837	7826	7815	7804	7793	7782	2	4	5	7	9
39	·7771	7760	7749	7738	7727	7716	7705	7694	7683	7672	2	4	6	7	9
40	·7660	7649	7638	7627	7615	7604	7593	7581	7570	7559	2	4	6	8	9
41	·7547	7536	7524	7513	7501	7490	7478	7466	7455	7443	2	4	6	8	10
42	·7431	7420	7408	7396	7385	7373	7361	7349	7337	7325	2	4	6	8	10
43	·7314	7302	7290	7278	7266	7254	7242	7230	7218	7206	2	4	6	8	10
44	·7193	7181	7169	7157	7145	7133	7120	7108	7096	7083	2	4	6	8	10

NATURAL COSINES

[Numbers in difference columns to be subtracted, not added.]

Degrees	0' 0°·0	6' 0°·1	12' 0°·2	18' 0°·3	24' 0°·4	30' 0°·5	36' 0°·6	42' 0°·7	48' 0°·8	54' 0°·9	Mean Differences 1 2 3	4 5
45	·7071	7059	7046	7034	7022	7009	6997	6984	6972	6959	2 4 6	8 10
46	·6947	6934	6921	6909	6896	6884	6871	6858	6845	6833	2 4 6	8 11
47	·6820	6807	6794	6782	6769	6756	6743	6730	6717	6704	2 4 6	9 11
48	·6691	6678	6665	6652	6639	6626	6613	6600	6587	6574	2 4 7	9 11
49	·6561	6547	6534	6521	6508	6494	6481	6468	6455	6441	2 4 7	9 11
50	·6428	6414	6401	6388	6374	6361	6347	6334	6320	6307	2 4 7	9 11
51	·6293	6280	6266	6252	6239	6225	6211	6198	6184	6170	2 5 7	9 11
52	·6157	6143	6129	6115	6101	6088	6074	6060	6046	6032	2 5 7	9 12
53	·6018	6004	5990	5976	5962	5948	5934	5920	5906	5892	2 5 7	9 12
54	·5878	5864	5850	5835	5821	5807	5793	5779	5764	5750	2 5 7	9 12
55	·5736	5721	5707	5693	5678	5664	5650	5635	5621	5606	2 5 7	10 12
56	·5592	5577	5563	5548	5534	5519	5505	5490	5476	5461	2 5 7	10 12
57	·5446	5432	5417	5402	5388	5373	5358	5344	5329	5314	2 5 7	10 12
58	·5299	5284	5270	5255	5240	5225	5210	5195	5180	5165	2 5 7	10 12
59	·5150	5135	5120	5105	5090	5075	5060	5045	5030	5015	3 5 8	10 13
60	·5000	4985	4970	4955	4939	4924	4909	4894	4879	4863	3 5 8	10 13
61	·4848	4833	4818	4802	4787	4772	4756	4741	4726	4710	3 5 8	10 13
62	·4695	4679	4664	4648	4633	4617	4602	4586	4571	4555	3 5 8	10 13
63	·4540	4524	4509	4493	4478	4462	4446	4431	4415	4399	3 5 8	10 13
64	·4384	4368	4352	4337	4321	4305	4289	4274	4258	4242	3 5 8	11 13
65	·4226	4210	4195	4179	4163	4147	4131	4115	4099	4083	3 5 8	11 13
66	·4067	4051	4035	4019	4003	3987	3971	3955	3939	3923	3 5 8	11 14
67	·3907	3891	3875	3859	3843	3827	3811	3795	3778	3762	3 5 8	11 14
68	·3746	3730	3714	3697	3681	3665	3649	3633	3616	3600	3 5 8	11 14
69	·3584	3567	3551	3535	3518	3502	3486	3469	3453	3437	3 5 8	11 14
70	·3420	3404	3387	3371	3355	3338	3322	3305	3289	3272	3 5 8	11 14
71	·3256	3239	3223	3206	3190	3173	3156	3140	3123	3107	3 6 8	11 14
72	·3090	3074	3057	3040	3024	3007	2990	2974	2957	2940	3 6 8	11 14
73	·2924	2907	2890	2874	2857	2840	2823	2807	2790	2773	3 6 8	11 14
74	·2756	2740	2723	2706	2689	2672	2656	2639	2622	2605	3 6 8	11 14
75	·2588	2571	2554	2538	2521	2504	2487	2470	2453	2436	3 6 8	11 14
76	·2419	2402	2385	2368	2351	2334	2317	2300	2284	2267	3 6 8	11 14
77	·2250	2233	2215	2198	2181	2164	2147	2130	2113	2096	3 6 9	11 14
78	·2079	2062	2045	2028	2011	1994	1977	1959	1942	1025	3 6 9	11 14
79	·1908	1891	1874	1857	1840	1822	1805	1788	1771	1754	3 6 9	11 14
80	·1736	1719	1702	1685	1668	1650	1633	1616	1599	1582	3 6 9	12 14
81	·1564	1547	1530	1513	1495	1478	1461	1444	1426	1409	3 6 9	12 14
82	·1392	1374	1357	1340	1323	1305	1288	1271	1253	1236	3 6 9	12 14
83	·1219	1201	1184	1167	1149	1132	1115	1097	1080	1063	3 6 9	12 14
84	·1045	1028	1011	0993	0976	0958	0941	0924	0906	0889	3 6 9	12 14
85	·0872	0854	0837	0819	0802	0785	0767	0750	0732	0715	3 6 9	12 15
86	·0698	0680	0663	0645	0628	0610	0593	0576	0558	0541	3 6 9	12 15
87	·0523	0506	0488	0471	0454	0436	0419	0401	0384	0366	3 6 9	12 15
88	·0349	0332	0314	0297	0279	0262	0244	0227	0209	0192	3 6 9	12 15
89	·0175	0157	0140	0122	0105	0087	0070	0052	0035	0017	3 6 9	12 15
90	·0000											

the correlation coefficient between temperature and hours of sunshine is approximately equal to 0.9.

Quite rightly, researchers get excited when they discover correlations as high as this between two variables, especially bearing in mind that positive correlation coefficients vary in size from as low as zero to one. However, as the scatterplots show, there is a significant degree of scatter or variability in the data, and the relationship between temperature and hours of sunshine is far from perfect. (Remember, if it were perfect, all the points would fall on a straight line.) In fact, the 'square of the correlation coefficient', usually referred to as R^2, gives a measure of the variability between the two variables which is accounted for by the correlation between them. In the present case, $(0.9)^2 \simeq 0.8$, so 0.8 or 80 per cent of the variability between temperatures and hours of sunshine in English south coast towns has been accounted for by the correlation between them. Put a different way, 20 per cent of the variability in the relationship between these two variables remains unexplained. In other words, knowing the hours of sunshine would not give us a precise way of predicting the temperature. If we wanted to get a better prediction of temperature we'd have to look at other factors that affect the temperature as well as the hours of sunshine. Nevertheless, that there *is* a strong relationship between temperature and hours of sunshine seems beyond reasonable doubt.

Another very helpful way of visualizing the correlation between two variables is to express the correlation geometrically. The correlation between two variables can in fact be expressed in terms of an angle between two straight lines. The straight lines, which are referred to as *vectors*, represent the variables and the angle between the lines represents the correlation. Vectors have two special qualities, namely, they represent the variables in *magnitude* and *direction*, relative to each other. Providing the variables are measured on the same scales, as is the case with standardized scores, the vectors that represent them will be of equal length; and, directionally, to represent the correlation coefficient, the angle between them is such that the cosine of the angle is numerically equal to the correlation coefficient. In our example in Fig. 1.6, to represent the correlation coefficient between temperature and hours of sunshine as vectors, the 'Temperature' axis and the 'Hours of sunshine' axis would be rotated until the angle between them was approximately equal to 25°, since reference to a table of cosines (Table 1.5) shows that the cosine of 25° is approximately equal to 0.9, the

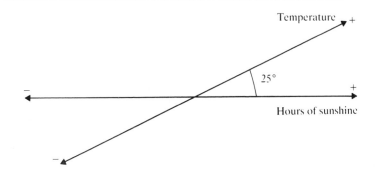

Figure 1.7   *Correlation between temperature and hours of sunshine represented by vectors.*

numerical value of the correlation coefficient between temperature and hours of sunshine in south coast towns.

Figure 1.7 shows the resultant diagram where the correlation between temperature and hours of sunshine is represented by vectors.

In the language of trigonometry, any angle between 0° and 180°, excluding 90°, is called *oblique*; any angle between 0° and 90° is called *acute*, and any angle between 90° and 180° is called *obtuse*. Acute and obtuse angles represent positive and negative correlations, respectively. For example, the acute angle of 60°, which gives a cosine of 0.5, represents a positive correlation of 0.5 and the obtuse angle of 120°, which gives a cosine of −0.5, represents a negative correlation of −0.5. (To refresh your memory of trigonometry, cosine 120° = −cosine (180° − 120° = −cosine 60°).

There are two particularly interesting instances when variables and their correlations are represented as vectors. One is when two variables are perfectly correlated (i.e. the correlation coefficient = 1), and the other is when they are completely uncorrelated (i.e. the correlation coefficient = 0). Reference to Table 1.5 indicates that the cosine of 0° = 1 and the cosine of 90° = 0. Thus, in the case of a perfect positive correlation, the angle between the two vectors would be 0° or, in other words, the two vectors would be superimposed, one on top of the other. Two completely uncorrelated variables are represented by two vectors at 90° to each other. In this instance, the variables are said to be *orthogonal*. Of course, this is just another way of saying they are uncorrelated, but it's a term worth remembering since it's often used in *factor analysis*, which we'll take a look at in Chapter 3.

As the angle between two vectors increases from 0° to 90°, the corresponding correlation between the two variables they represent decreases from 1 to 0, as an inspection of Table 1.5 shows.

If the foregoing discussion of vectors and cosines seems a little obscure, its relevance will become immediately apparent when factor analysis is discussed in Chapter 3.

Before leaving this section, a few words are appropriate about what the correlation between two variables does *not* tell us. A correlation coefficient on its own gives us no information whatsoever about the *reasons* why the variables are related; all it tells us is that they *are* related. In the example we have considered here, it might be tempting to argue that sunshine *causes* an increase in temperature, and this may well be a plausible explanation (although anyone who has been in Canada on a sunny January day might also argue that sunshine causes the temperature to be very low!). The tendency to identify 'correlation' and 'causation' is so widespread that it is worth while to point out the danger of doing so with a striking example.

Consider the sorts of figures produced by the Statistical Bureau of the EC that relate *per capita* strike days to increases in the Gross National Product (GNP) *per capita* and to the rate of annual inflation. We can see in Table 1.6 that there are two rather striking relationships. There is a negative correlation between '*per capita* strike days' and 'increase in the *per capita* GNP'; and a positive correlation between '*per capita* strike days' and 'rate of annual inflation'. What conclusions can we draw from these figures?

Table 1.6
Per capita strike days, GNP and the rate of inflation

Country	Strike days per 100 workers 1969–1974	Increase of the national product per capita in pounds sterling 1968–1973	Rate of inflation (annual) average %
West Germany	240	1193	5.2
France	901	838	7.4
UK	9035	436	8.9
Italy	5083	389	8.0

During the seventies, politicians of a particular bent had no doubt at all how these figures were to be interpreted. They claimed that the figures demonstrated that strikes, and, by association, the trade union movement, were the biggest *cause* of decline in the United Kingdom. (The point being discussed here is not whether such an argument is 'correct' in itself, but whether the data in Table 1.6 'prove' the truth of the argument.) The competent researcher would point out that there are at least two things wrong with drawing this conclusion from Table 1.6. Firstly, the figures are comparing apples with pears, so to speak. Strike days per 100 workers is related to the GNP of the different countries as a *whole*. It seems more reasonable to assume that most strikes are not called by union leaders to decrease the GNP but to improve the living standards of the workers they represent; and it would be interesting to see whether they did so. Secondly, the inference from correlation to causation, that is, that strikes *caused* a decrease in the GNP and *caused* an increase in the rate of inflation is very dubious reasoning, although as a species of argument it is very common indeed. This, in fact, is a very basic warning about the misuse of statistics and it would be very familiar to competent professional researchers, who are well versed in the nature and problems of statistical inference. In addition, the sensitive researcher would seek as many explanations as possible of the observed relationships to attempt to find the one that best seems to fit *all* the available data in the most *consistent* manner. Being sensitive in this way is partly like being a good detective, who searches diligently and meticulously for as many clues as possible and who does not reach a hasty conclusion. Skills of advocacy, as well as those of detection, are a necessary part of this sensitivity; the competent researcher, having arrived at a conclusion, will subject it to the most searching cross-examination until convinced that the judgement made is the best that can be reached in the circumstances. In the present example, the researcher may suggest that strikes are yet another manifestation of the dissatisfaction and discontent that pervade a society that is not dealing very well with its inefficiencies and inequalities. In this case, it would be necessary to search for further evidence to put this idea to the test.

Errors in reasoning of the sort we've just discussed are more frequent than we would like to believe. A common example in advertising is the one that goes something like this: 'When you correlate the amount of advertising for a product, you often find that large advertising budgets *cause* increased sales.' Much as advertising agencies might like us all to believe this, it's not nearly as clear-cut as might be made out. It is always possible that increased sales provide budget surpluses that cause greater advertising expenditure. Usually, it is not immediately obvious what causes what!

The limitations of statistics

Written over the door to Plato's Academy were the words: 'Let no one enter here who is ignorant of geometry'. If we were to think of a similar suitable inscription to hang over the door of a survey research company it might be something like: 'Let no one enter here who is (completely) ignorant of statistics'. Statistical ideas are at the root of systematic survey research, and, notwithstanding Disraeli's famous aphorism concerning their mendacity, it is difficult to conceive of a competent researcher who is not familiar with some of the techniques that have arisen from statistics.

However, the instances in which we can rely solely on statistical inferences in research are extremely rare and we should not be blinded or impressed by high-falutin mathematics. The important thing to understand about statistical analyses is that results arising from their application apply *only* to the conditions and considerations that prevailed at the time of analysis. Judgements about their relevance to the real world can only be made sensibly by people who have a good grasp of the total situation. Thus, their applicability depends on sound judgement and not upon statistical theory; they depend on the *sensitivity* of the researcher. It is worth spelling out the limitations of statistical inference in a little more detail because they are most frequently overlooked.

First, they refer only to the populations studied and the conditions under which they were studied. In a strict sense, the results in our example of weather conditions in English south coast towns apply only to those towns on the autumn day in question. This means, for example, that any inferences we might draw about the weather conditions on the south coast on a *particular* autumn day cannot, logically, apply to any other day, let alone any towns in a different locality.

Secondly, inferences refer only to the techniques and methods used in the study. For example, inferences from the results of a particular survey using a questionnaire administered in a face-to-face interview, do not apply to a different but closely related questionnaire, administered over a telephone.

In many instances in survey research, these limitations may not have serious practical consequences and broadly similar results have been obtained in somewhat different conditions and circumstances. Decisions can usually be made more wisely on the basis of sound research, warts and all. But decisions based on unsound research are unlikely to be wiser than those based on sound experience. Really wise decisions are those based on sound research and sound experience!

2. Introduction to the data set

Background

In the chapters that follow this one, various methods of analysis are described. Here, we'll look at the data that crop up throughout the rest of the book which are used to explain the different methods of analysis. These data arose from a research project concerned with things called 'personal alarms'.

Personal alarms are small devices usually carried on the person or, in the case of women, in the handbag. When activated, personal alarms emit a loud and distinctive noise, and the purpose of their doing so is primarily to alarm a would-be attacker so that he runs off. A secondary purpose is to alert passers-by in the (often vain) hope that they will come to the victim's assistance.

Personal alarms are not seen as deterrents since most street attacks are based on surprise, usually from behind. In these circumstances, the assailant is hardly likely to be scrutinizing his potential victim closely enough to spot whether she is carrying a personal alarm. Moreover, most attacks take place at night in poorly lit areas.

Personal alarms are supposed to give their carriers confidence but not *so* much confidence that they do not feel at risk. Indeed, some experts, such as crime prevention officers, feel that personal alarms are only one small measure that potential victims should take to reduce the risk of being attacked, while others do not recommend the use of personal alarms at all since they feel they might even give rise to unnecessary anxieties about safety.

Nevertheless, as we all know, street crimes such as rape and mugging are increasing at an alarming rate and as a consequence the market for personal alarms is also increasing. Our client, therefore, was interested in entering the market especially since there were some strong indications at the time that the personal alarms available could be improved upon. Accordingly, a design company was commissioned to produce a number of models with various features which were thought to be improvements on those already on the market. Four such models were produced. As part of a larger investigation, research was undertaken in which the four models were evaluated by a sample of people in the target market for personal alarms.

The target market was defined by reference to the *British Crime Survey*, produced annually, which consistently shows that the people fearing most for their personal safety are women, the elderly and residents of inner-city areas. A sample of 100 people in this target group took part in the design research project.

The interview

The interview began with a set of questions designed to discover the extent to which respondents were concerned about the possibility of themselves and members of their families being victims of a street attack; the circumstances in which they were most fearful; whether they or other members of their families and their friends took precautions against street attacks; and, if so, what these precautions were, with particular reference to personal alarms and their usefulness.

During the next part of the interview—the part that generated the data that will concern us most—respondents evaluated the four models developed by the design company. To place their evaluations in a realistic context, the four models were evaluated along with four personal alarms actually on the market, so respondents evaluated eight models in total.

One of the purposes of the evaluations was to discover which of the eight models were perceived to be similar to each other. Our reason for wishing to know this was to see how successful the designers had been in creating something different. This was achieved by showing respondents all eight models and encouraging them to handle the alarms. The models were then presented to respondents in pairs and the respondents' task was to judge how similar the members of each pair were to each other. If a respondent felt the members of a pair were very similar to each other, they were instructed to give a score of '10'. If they felt the members of a pair were very dissimilar to each other, they were instructed to give a score of '0'. They were told that they could give any score between '0' and '10', depending on how similar or dissimilar they thought the members of each pair were to each other. In this manner, each respondent judged the similarity of the members of all possible pairs of models.

Following this, respondents were asked about each model individually. The interviewer read out 15 statements and after each statement respondents were asked to say whether they thought the statement described the model in question. These were statements which earlier research had suggested were important in determining people's overall evaluation of alarms. They were:

1. Could be set off almost as a reflex action
2. Would be difficult for an attacker to switch off
3. Could be carried to be very handy when needed
4. Would be difficult to break
5. Feels comfortable in the hand
6. Could be easily kept in the pocket
7. Would be difficult for an attacker to take it off me
8. I could keep a firm grip of it, if attacked
9. Solidly built
10. An attacker might have second thoughts about attacking me if he saw me with it
11. Would fit easily into a handbag
12. Could be easily worn on the person
13. An attacker might be frightened that I might attack him with it
14. Looks as if it would give off a very loud noise
15. I would be embarrassed to carry it round with me

After all eight models had been evaluated in this way, respondents were invited to evaluate the statements themselves; that is, they were asked to say which of the statements described something 'good' for a personal alarm to be or to have. Finally, the design evaluation concluded with respondents choosing which one of the eight models they liked best. The remainder of the questionnaire was devoted to collecting data about the respondents' price expectations of each of the models, together with personal details such as age and occupation.

Preliminary data analysis: the similarity–dissimilarity matrix

Table 2.1 shows the average similarity scores between the eight
models that have been labelled A to H. (For interest, the 'new'
models were E, F, G and H and the models already on the market
were A, B, C and D.)

The scores in Table 2.1 have been calculated from the individual
scores that were given when respondents were asked to give each
pair of models a score between '0' and '10' according to how similar
they thought they were to each other.

This matrix is made up of a lot of figures like the charts of
distances between towns that are found, for example, in the *Automobile Association Handbook*. Of course, in this case, an entry in
the matrix, for example, where the 'B' column intercepts the 'C'
row, does not show the distance between model B and model C but,
rather, the overall *similarity* between these two models, as perceived by a sample of the target market for personal alarms. Where
the 'B' column meets the 'B' row, the 'C' column the 'C' row, and so
on, an 'X' has been inserted to indicate identity. Also notice that the
upper half of the matrix (above the 'X' diagonal) is the mirror image
of the lower half, so all the information in Table 2.1 is contained
within one half of the matrix.

It is difficult for anyone who is not an arithmetical genius to get
much sense out of Table 2.1 merely by inspection, but it is possible

Table 2.1
Similarity matrix

	Models							
	A	B	C	D	E	F	G	H
A	X	8.2	7.1	7.2	4.2	6.7	5.5	8.3
B	8.2	X	5.7	5.5	3.7	5.7	5.5	8.2
C	7.1	5.7	X	9.0	1.6	4.5	2.7	5.3
D	7.2	5.5	9.0	X	2.2	5.3	3.4	5.6
E	4.2	3.7	1.6	2.2	X	7.2	8.4	5.4
F	6.7	5.7	4.5	5.3	7.2	X	8.0	7.5
G	5.5	5.5	2.7	3.4	8.4	8.0	X	6.9
H	8.3	8.2	5.3	5.6	5.4	7.5	6.9	X

Table 2.2
Simplified similarity matrix

	Models							
	A	B	C	D	E	F	G	H
A	10	8	7	7	4	7	6	8
B	8	10	6	6	4	6	6	8
C	7	6	10	9	2	5	3	5
D	7	6	9	10	2	5	3	6
E	4	4	2	2	10	7	8	5
F	7	6	5	5	7	10	8	8
G	6	6	3	3	8	8	10	7
H	8	8	5	6	5	8	7	10

to make the figures more comprehensive by 'simplifying' them and juggling them about a bit.

One fairly basic step towards making the figures more comprehensible is to forget about being really accurate and to simplify them in an acceptable way. For example, we can begin by dropping the decimal place and 'rounding up' or 'rounding down' as appropriate. This gives us a new set of figures, shown in Table 2.2. (In this table, we have also replaced the Xs by 10s to indicate perfect similarity or identity.) It has to be admitted that this simplification has not made the figures much easier to comprehend and we have also lost some information—by rounding the figures, we have made them less accurate. One way of making the 'meaning' of the figures easier to grasp is to rearrange or juggle them so as to bring out any 'natural' groups that may be there. There are several ways of doing this, each with its own advantages and disadvantages, and some have more visual impact than others.

Before making an attempt to give the matrix in Table 2.2 more of a visual impact, we can try to rearrange the rows and columns in such a way as to produce a more comprehensible pattern. For instance, the arrangement in Table 2.3 produces a matrix in which the rows and columns have been juggled to show more clearly the groups of models that are similar to each other, and a square has been placed round each group.

The top left of the matrix in Table 2.3 is occupied by two models, C and D, which respondents clearly perceived to be similar to each other. The bottom right of the matrix is occupied by models G and E, which are also perceived to be similar to each other. In the middle of the matrix, models A, B and H also form a group of models which seem to be perceived as being similar to each other. However, model F is not so clear-cut. We could just as reasonably place model F with either model H or model G, so perhaps we ought to leave it on its own. This arrangement does seem to highlight the overall structure of the respondents' perceptions of the similarities between the eight models. Moreover, the rearranged matrix in Table 2.3 also shows that the group or cluster containing models C and D is quite dissimilar from the cluster containing models G and E, and the cluster containing models A, B and H is 'half-way' between these two clusters.

One visually appealing way of representing the rearranged matrix in Table 2.3 is to replace the actual figures by dots, tones, or lines, i.e. by shadings to represent the degree of similarity between the models. One way of doing this is given in Fig. 2.1, where the pattern of shading produces dark and light areas, with the dark areas representing degrees of similarity between the models and the light areas representing degrees of dissimilarity.

The trouble with all of these procedures is that it is not always

Table 2.3
Simplified similarity matrix with rows and columns reordered models

Models								
	C	D	A	B	H	F	G	E
C	10	9	7	6	6	5	3	2
D	9	10	7	6	5	5	3	2
A	7	7	10	8	8	7	6	4
B	6	6	8	10	8	6	6	4
H	5	6	8	8	10	8	7	5
F	5	5	7	6	8	10	8	7
G	3	3	6	6	7	8	10	8
E	2	2	4	4	5	7	8	10

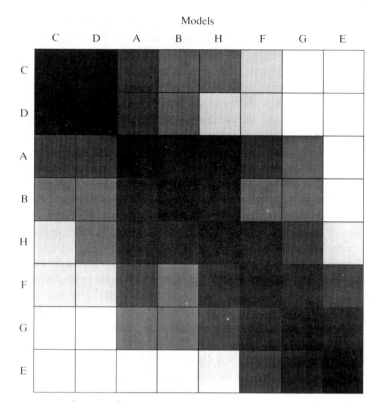

Figure 2.1 *Shaded similarity matrix.*

obvious how to arrange the rows and columns to produce the most
clear-cut patterns, although computer programmes are now available which take the 'slog' out of it for you. The important point to
take from all this, however, is that you do not need to be a
mathematical wizard to begin to see sensible patterns in what at first
glance looks like a complicated and forbidding set of figures. In
Chapter 4, concerned with mapping, we shall see a much more
intellectually appealing and elegant way of achieving a clear picture
of the relationships of similarity and dissimilarity between the
models.

Another way to exhibit the perceived similarities between the
models in a visual form is to turn them into what is known as a 'tree
diagram' or 'dendrogram'. This has the effect of showing the perceived structure of similarities between the models as a modified

hierarchy, with some gaps and jumps. We will now look at how we might construct a dendrogram from the figures in Table 2.3, and it is worth noting in passing that dendrograms feature prominently in cluster analysis methods, which are discussed in Chapter 5.

To construct a dendrogram from Table 2.1, we will need to develop some rules. Let's begin by grouping together the two models most similar to each other. At the highest level of similarity, S, we find C and D with $S = 9$, so we will join these together at a *level* of $S = 9$. We will now look for the next highest S; this is between E and G with $S = 8.4$, so E and G are joined together at level $S = 8.4$. Proceeding in this fashion A and H would be joined together at level $S = 8.3$. The next highest level is $S = 8.2$, between B and A, and between B and H. We, therefore, join B with A and H at level $S = 8.2$. (Note that A and H have already been joined together at level $S = 8.3$.) The next highest S is 8.0, between F and G, but G has already been joined to E at level $S = 8.4$. Let's therefore invoke a rule to join F with both E and G at level $S = 8.0$. In general terms this rule can be stated as, 'any new model for membership of a group can be joined to an existing group on the basis of the highest level of similarity of any member of the existing group'. That is, only a 'single link' is required between two models for them to merge. The next highest level of S is 7.2, between A and D. Again applying our *single-linkage* rule, we will join the 'already formed group' containing C and D at level $S = 7.2$. The next highest level of S is 6.9 between G and H. Applying our single-linkage rule yet again, we join all the existing groups together at level $S = 6.9$. The dendrogram representing the results of joining together the models using a single-linkage rule is shown in Fig. 2.2. Each step where a pair of models (or group of models) is merged or joined is represented as a branch in the tree. The tree or dendrogram represents a *hierarchical* organization of the relations between the eight models. At the lowest level, all eight models are independent, and at successive levels models are merged until, finally, at the highest level, all eight models are joined into one group. The way we interpret the dendrogram in Fig. 2.2 is as follows. For example, if we want to know how similar model F is to level C, we have to discover to what level of the tree we have to go before we find a bridge to get from one to the other. In this example, we have to go to the highest level of the tree which is the lowest level of similarity. We should note, however, that by applying our single-linkage rule we have lost some of the information concerning the similarities between the eight

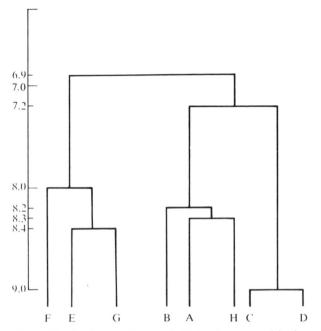

Figure 2.2 *Dendrogram showing similarities between models (drawn from data in Table 2.1).*

models. For example, from Table 2.1, the similarity between models C and E is given as $S = 1.6$, yet in the dendrogram the similarity between C and E is indicated to be $S = 6.9$. This is because the merging process of single-linkage fails to take account of all the various similarities in Table 2.1. There are other linkage rules which we'll consider in Chapter 5, on cluster analysis, which are more sensitive to the different similarity scores in Table 2.1 and which accordingly tend to produce different trees. Nevertheless, the dendrogram in Fig. 2.2 has certainly suggested that we have three clusters of models, namely E, F and G; A, B and H; and C and D. Moreover, it tends to confirm the groups or clusters suggested in Table 2.3. The exception seems to be model F which has been clustered with models E and G in Fig. 2.2 although it was not as closely related to them in Table 2.3.

In the questionnaire we also asked our sample of respondents to volunteer which of 15 statements best described each of the models. Table 2.4 shows the percentages of respondents who thought which statements described which models. The columns of Table 2.4 have been ordered in line with those in Table 2.3 but in addition there is a column labelled 'Ideal'. This is a description we have given to the percentages of respondents who thought each of the statements described something that was 'good' for a personal alarm to be or to have. The 'Ideal' model does not exist, of course, but it can be thought of as an evaluative component against which each of the real models can be judged. The figures in Table 2.4 have been simplified by rounding to the nearest ten.

In addition, the rows in the table have been arranged in descending order of magnitude from those most highly evaluated to those least highly evaluated, that is, from the highest to the lowest percentages on the Ideal.

Ignoring for the moment the marginal totals in Table 2.4, if we arbitrarily select the most important attributes of a personal alarm to be those characteristics selected by 70 per cent (or more) of the respondents as something 'good' for a personal alarm to be or to have, the most highly evaluated characteristics of a personal alarm are:

1. Could be set off almost as a reflex action
2. Would be difficult for an attacker to switch off
3. Could be carried to be very handy when needed
4. Would be difficult to break
5. Feels comfortable in the hand
6. Could be easily kept in the pocket
7. Would be difficult for an attacker to take it off me
8. I could keep a firm grip of it, if attacked

Also, by applying our arbitrary '70 per cent rule', we can see that all the models except H achieve 'less than ideal' ratings on two or more of these important eight statements. More specifically, each of the models is inadequate in the following respects:

Model C:
Would *not* be difficult for an attacker to switch off
Would *not* be difficult to break
Would *not* be difficult for an attacker to take it off me
I could *not* keep a firm grip of it, if attacked.
Model D:
Would *not* be difficult for an attacker to switch off
Would *not* be difficult to break
Would *not* be difficult for an attacker to take it off me
Model A:
Would *not* be difficult to break

Table 2.4
Percentages of respondents who thought which statements described which models and the Ideal

Statements	Models								Ideal	Total
	C	D	A	B	H	F	G	E		
1. Could be set off almost as a reflex action	80	80	80	100	90	80	60	30	90	690
2. Would be difficult for an attacker to switch off	0	0	100	100	100	100	100	30	90	620
3. Could be carried to be very handy when needed	90	90	80	80	90	40	50	10	80	610
4. Would be difficult to break	10	50	50	30	90	100	100	100	80	610
5. Feels comfortable in the hand	100	100	90	90	90	30	0	0	80	580
6. Could be easily kept in the pocket	100	100	90	60	80	70	10	70	70	650
7. Would be difficult for an attacker to take it off me	40	50	40	70	70	10	40	0	70	390
8. I could keep a firm grip of it, if attacked	60	70	60	80	90	20	60	0	70	510

										Total
9. Solidly built	40	70	80	60	100	100	100	100	60	710
10. An attacker might have second thoughts about attacking me if he saw me with it	20	20	40	50	50	40	60	40	60	380
11. Could fit easily into a handbag	100	100	90	90	90	80	40	10	50	650
12. Could be easily worn on the person	10	10	50	70	80	0	0	0	50	270
13. An attacker might be frightened that I might attack him with it	0	0	20	30	60	40	80	80	40	350
14. Looks as if it would give off a very loud noise	20	20	40	70	40	50	70	70	40	420
15. I would be embarrassed to carry it round with me	10	0	10	20	40	30	50	80	0	240
Total	680	760	920	1000	1160	790	820	620	930	7680 Grand total

Would *not* be difficult for attacker to take it off me
I could *not* keep a firm grip of it, if attacked
Model B:
Would *not* be difficult to break
Could *not* be easily kept in the pocket
Model H:
No inadequacies
Model F:
Could *not* be carried to be very handy when needed
Does *not* feel comfortable in the hand
Would *not* be difficult for an attacker to take it off me
I could *not* keep a firm grip of it, if attacked
Model G:
Could *not* be set off as a reflex action
Could *not* be carried to be very handy when needed
Does *not* feel comfortable in the hand
Could *not* be easily kept in the pocket
Would *not* be difficult for an attacker to take it off me
I could *not* keep a firm grip of it, if attacked
Model E:
Could *not* be set off as a reflex action
Would *not* be difficult for an attacker to switch off
Could *not* be carried to be very handy when needed
Does *not* feel comfortable in the hand
Would *not* be very difficult for an attacker to take it off me
I could *not* keep a firm grip of it, if attacked

It is also possible to gain further insight into the data in Table 2.4 by examining the marginal totals. For example, model H has twice as many endorsements overall (1160) as model E (620); and statement 9 has three times the level of endorsement across all the models (710) in comparison with statement 15 (240). The variability in the row and column totals in a table like Table 2.4 gives rise to what are known as *row effects* and *column effects*. These effects can make interpretation of the relative differences between the models on the statements more tricky than it might seem by simply analysing the differences between the models on each of the statements one at a time. For example, it is all very well to note that whereas 90 per cent of respondents thought model H 'could be set off almost as a reflex action' (statement 1), only 30 per cent of respondents thought this of model E, but how dramatic or significant is this

difference of sixty per cent relative to the general level of endorsement of all the statements on all the models? One way of making
an attempt to answer this question would be by removing the
row effects and the column effects from Table 2.4. Let's try to do
this.

First of all we'll try to remove the row effects. This is achieved by
taking an average of the values in each row, such as the mean or the
median, and subtracting these values from the corresponding rows
of actual or observed values. This will leave a table of *row residuals*.

In our case, since the various row values are far from normally
distributed, the median is a better measure of average row values
than the mean. If you remember, in Chapter 1 it was indicated that
the median value of a series of numbers is the number in the series
above and below which 50 per cent of all the other numbers lie. So,
for example, in the first row of Table 2.4 (ignoring, of course the
'total' figure), the median value is 80. In the second row the median
value is 100 and in the third row it is 80. And so on. The median
values for each of the 15 rows are, in order:

$$80, 100, 80, 85, 90, 75, 45, 65, 90, 45, 90, 30, 40, 40, 25$$

Now, to remove the row effects from Table 2.4, we subtract from
each value the median for that row. This gives rise to Table 2.5, a
table of row residuals.

To remove the column effects, we repeat the procedure, only this
time we calculate differences from the column averages. Again,
we'll take the median value as our measure of the average. We first
of all find the median of each of the ten columns of Table 2.5
(including the column of row medians). These are, from left to
right:

$$0, 0, 0, 0, 10, 0, 0, -10, 0, 70$$

We now subtract these from their respective column values to give
us a second table of residuals and table effects. The result is shown
in Table 2.6. The analysis we have just carried out is known as
two-way analysis.

The table effects have three components—*column effects*, *row
effects* and a *common value*. The first thing to note is that the table
effects are additive. To produce the original table of values—Table
2.4—we add four numbers for each cell. For example, to compute
the percentage of respondents who think statement 1 applies to

Table 2.5
Table of row residuals (from Table 2.4)

Statements	Models								Ideal	Row median
	C	D	A	B	H	F	G	E		
1. Could be set off almost as a reflex action	0	0	0	20	10	0	−20	−50	10	80
2. Would be difficult for an attacker to switch off	−100	−100	0	0	0	0	0	−70	−10	100
3. Could be carried to be very handy when needed	10	10	0	0	10	−40	−30	−70	0	80
4. Would be difficult to break	−70	−30	−30	−50	10	20	20	20	0	80
5. Feels comfortable in the hand	10	10	0	0	0	−60	−90	−90	−10	90
6. Could be easily kept in the pocket	30	30	20	−10	10	0	−60	0	0	70
7. Would be difficult for an attacker to take it off me	0	10	0	30	30	−30	0	−40	30	40

8. I could keep a firm grip of it, if attacked	0	10	0	20	30	−40	0	−60	10	60
9. Solidly built	−40	−10	0	−20	20	20	20	20	−20	80
10. An attacker might have second thoughts about attacking me if he saw me with it	−20	−20	10	10	0	0	20	0	20	40
11. Could fit easily into a handbag	10	10	0	0	0	−10	−50	−80	−40	90
12. Could be easily worn on the person	0	0	40	60	70	−10	−10	−10	40	10
13. An attacker might be frightened that I might attack him with it	−40	−40	−20	−10	20	0	40	40	0	40
14. Looks as if it would give off a very loud noise	−20	−20	0	30	0	10	30	30	0	40
15. I would be embarrassed to carry it round with me	−10	−20	−10	0	20	10	30	60	−20	20

Table 2.6
Residuals and table effects

Statements	Models								Ideal	Row effects
	C	D	A	B	H	F	G	E		
1. Could be set off almost as a reflex action	0	0	0	20	0	0	-20	-40	10	10
2. Would be difficult for an attacker to switch off	-100	-100	0	0	-10	0	0	-60	-10	30
3. Could be carried to be very handy when needed	10	10	0	0	0	-40	-30	-60	0	10
4. Would be difficult to break	-70	-30	-30	-50	0	20	20	30	0	10
5. Feels comfortable in the hand	10	10	0	0	-10	-60	-90	-80	-10	20
6. Could be easily kept in the pocket	30	30	20	-10	0	0	-60	10	0	0
7. Would be difficult for an attacker to take it off me	0	10	0	30	20	-30	0	-30	30	-30

8. I could keep a firm grip of it, if attacked	0	10	0	0	20	20	-40	0	-50	10	-10
9. Solidly built	-40	-10	-20	0	-20	10	20	20	30	-20	10
10. An attacker might have second thoughts about attacking me if he saw me with it	-20	-20	10	0	10	0	0	20	10	20	-30
11. Could fit easily into a handbag	10	10	10	0	0	-10	-10	-50	-70	-40	20
12. Could be easily worn on the person	0	0	0	40	60	60	-10	-10	0	40	-60
13. An attacker might be frightened that I might attack him with it	-40	-40	-20	-10	-10	10	0	40	50	0	-30
14. Looks as if it would give off a very loud noise	-20	-20	0	0	30	-10	10	30	40	0	-30
15. I would be embarrassed to carry it round with me	-10	-20	-20	-10	0	10	10	30	70	-20	-30
Column effects	0	0	0	0	0	10	0	0	-10	0	70= Common value

43

model C, we add the common value (70), the row effect (10), the column effect (0) and the residual (0):

$$70 + 10 + 0 + 0 = 80$$

What the common value tells us is that 'on average' a statement was thought to apply to a model by 70 per cent of respondents. The column effects tell us how much, taken over all the statements, each model was above or below 'the average' (common value); and the 'row effect' tells us how much, taken over all the models, each statement was above or below 'the average' or common value. The residuals tell us what is left, so to speak, when the row effects, the column effects and the common value have been removed from the original values in Table 2.4.

In a two-way analysis of the sort we've just carried out, if the residuals are small or show no obvious pattern then the analysis is complete and our derived table (i.e. Table 2.6) has been decomposed adequately into its constituent parts—residuals, row effects, column effects and the common value. The residuals in Table 2.6 could hardly be described as small since they range from 100 to −90. However, they do not exhibit any strong patterns in the sense that they are approximately randomly or normally distributed with a mean of approximately 0 and a standard deviation of approximately 30. Thus, our two-way analysis is complete.

In many instances where two-way analyses are carried out where the residuals approximate to zero, much of the interest focuses on an examination of the row and column effects. In our case, however, the residuals are more interesting than the table effects since they show us which statements apply particularly to which models when the table effects have been removed. In other words, they show which statements *discriminate* between the models. Moreover, as an examination of the residuals shows, statements can discriminate in a *positive* sense and in a *negative* sense (shown by the minus signs in Table 2.6). So, for example, statement 15—'I would be embarrassed to carry it round with me'—applies particularly to model E, relative to all other statements and models, and it does *not* apply to the Ideal relative to all other statements and models.

Of particular interest are those statements that discriminate the Ideal from the eight models. Clearly these will suggest the kinds of qualities that a successful personal alarm should possess or *not* possess. So, for example, statement 7—'would be difficult for an attacker to take it off me'—describes a quality that a personal alarm should possess. We can see that models B and H both seem to have this quality and that models F and E lack it. Moreover, from Table 2.6 we can also see which of the statements discriminate most between the models. Statement 1, for example, shows relatively little discrimination between the models.

Table 2.6 seems to provide a much clearer picture of the differences between the models than Table 2.4. At the same time it does not lose any information since by adding the residuals to the common value and to their row and column effects it is possible to reproduce Table 2.4 from Table 2.6.

Another way of removing the row and column effects from Table 2.4 can be achieved if we subtract from each of the cell values in Table 2.4 the value we might 'expect' to find if there were no row or column effects.

For example, the expected value for model C on statement 1 is found by multiplying the appropriate marginal totals together and dividing this product by the grand total, i.e.

$$\frac{690 \times 680}{7680} = 61$$

We then subtract this from the observed value, 80. This gives a value of 19.

Similarly, the expected value for model E on statement 15 is found as follows:

$$\frac{620 \times 240}{7680} = 19$$

When we subtract this from the observed value, again 80, this gives us 61. Table 2.7 shows the differences between the observed and expected values for Table 2.4, calculated in this manner. The principles involved in compiling Table 2.7 are the basis of a much used statistical technique known as *chi-squared* analysis. The notion behind this analysis is that if there were no row effects and no column effects, then the difference between the observed value and the expected value would equal zero. The data in Table 2.7 clearly show that there are row and column effects but, unlike the data in Table 2.6, they do not indicate the magnitude of the effects due to the rows and columns separately. Moreover, it is difficult to see the relationship between the data in Tables 2.4 and 2.7. This is not so with Tables 2.4 and 2.6, where it is an easy matter to convert the

Table 2.7
Differences between observed and expected values in Table 2.4

Statements	Models								Ideal
	C	D	A	B	H	F	G	E	
1. Could be set off almost as a reflex action	19	12	−3	10	−14	9	−14	−26	6
2. Would be difficult for an attacker to switch off	−55	−61	26	19	6	36	34	−20	15
3. Could be carried to be very handy when needed	36	30	7	1	−2	−23	−15	−39	6
4. Would be difficult to break	−44	−10	−23	−49	−2	37	35	51	6
5. Feels comfortable in the hand	49	43	21	14	2	−30	−62	−47	10
6. Could be easily kept in the pocket	42	36	12	−25	−18	3	−59	18	−9
7. Would be difficult for an attacker to take it off me	5	11	−7	19	11	−30	−2	−31	23
8. I could keep a firm grip of it, if attacked	15	20	−1	14	13	−32	6	−41	8
9. Solidly built	−23	0	−5	−32	−7	27	24	42	−26
10. An attacker might have second thoughts about attacking me if he saw me with it	−14	−18	−6	1	−7	1	19	9	14
11. Could fit easily into a handbag	42	36	12	5	−8	13	−29	−42	−29
12. Could be easily worn on the person	−14	−17	18	35	39	−28	−29	−22	17
13. An attacker might be frightened that I might attack him with it	−31	−35	−22	−16	7	4	43	52	−2
14. Looks as if it would give off a very loud noise	−17	−22	−10	15	−23	7	25	36	−11
15. I would be embarrassed to carry it round with me	−11	−24	−19	−11	4	5	24	61	−29

latter to the former. Overall, two-way analysis seems more useful to us than chi-squared analysis.

During the interview, we also asked respondents which of the eight models they liked best. Model H was chosen by more respondents than any other model and model E was not chosen by a single person. From Tables 2.4 and 2.6, it can be clearly seen that the profile of model H is the most similar to the Ideal and the profile of model E is the least similar.

The preliminary analyses carried out on the data in this chapter have given us some insight into how the eight models were perceived and evaluated. However, these analyses are quite time-consuming, and it is difficult to imagine that a busy researcher will always have the time to carry out some of the painstaking tasks that we've done here. Nevertheless, by now we should all be familiar with the data and be ready to see how various methods of analysis can be used to make our understanding of them more comprehensive and easier.

3. Principal components analysis and factor analysis

Background

The 'names' principal components analysis and factor analysis are frequently used in a fairly loose fashion in survey research to the extent that when researchers have actually carried out a principal components analysis they often report that they have carried out a factor analysis. The reason for this is that the two procedures are quite similar or, put another way, the differences between them are not immediately obvious. Indeed, many research practitioners who attempt an explanation of the differences between the two methods often end up getting it wrong. In this chapter, the emphasis will be on principal components analysis, which is more widely used in survey research than factor analysis. Moreover, since an explanation of the differences between factor analysis and principal components analysis is rather technical, no attempt at one will be made here. Indeed, the terms factor analysis and principal components analysis will be used as if they were interchangeable even though, in fact, they are not.

Factor analysis has a long and chequered history. In psychology, for example, its application has been in areas of acute political sensitivity such as the study of the nature and measurement of intelligence. Fortunately, from the viewpoint of this chapter, its use in survey research has usually been empirically based rather than theoretical, so none of the intense political arguments have surrounded this application. Nevertheless, its 'non-political' nature in survey research does not mean factor analysis has been free from controversy or polemic. On the contrary, it has probably generated more heat than any other method of analysis. The reason for this is that factor analysis is without doubt one of the most abused techniques in survey research.

The purpose of this chapter is to explain how factor analysis can be used without 'too much abuse' in its typical mode of application in survey research, namely, in data reduction. Its use in data reduction can best be explained by example. It is fairly common in survey research for objects such as products or political parties to be rated on large numbers of attitude statements that are deemed by the researchers to be relevant to an individual's choice in some way. Typically, the number of these statements varies between approximately 15 and 60. They are usually developed through knowledge or experience of the markets or topics under study, or they may be generated by exploratory, qualitative research. (However, how they have arisen and whether they are relevant need not concern us.) Respondents are usually asked to indicate whether they 'agree' or 'disagree' that the statements apply to the objects in question. Sometimes a rating scale is used (for example, where 'agree strongly' might be given a 'score' of '5', 'agree' a 'score' of '4', 'undecided' a 'score' of '3', 'disagree' a 'score' of '2' and 'disagree strongly' a 'score' of '1'). On other occasions respondents might be asked to indicate simply whether the statements in question apply to the object in question. In the latter case, which was the procedure used in the personal alarm study, the scoring system is '1' if a statement was thought to apply to a model and '0' if the statement was not thought to apply.

These procedures can give rise to a large amount of data and, rather than comparing products or political parties on 60 or so individual statements, survey researchers might be tempted to apply factor analysis to the data to see if the 60 or so statements can be reduced to a smaller number of 'factors' without 'losing too much information'.

Perhaps the foregoing discussion can be brought to life by a simple example. In Chapter 1, the correlation coefficient between 'temperature' and 'hours of sunshine' in English south coast towns was calculated to be approximately equal to 0.9. Moreover, it was suggested that the correlation coefficient between these two variables could be represented by the angle between two variables when they are represented by vectors (see Fig. 1.7). The question that factor analysis might attempt to answer is:

Can these two vectors be replaced by a single *reference vector*, known as a factor, such that the factor retains most of the information concerning the correlation between the original two variables?

Let's now try to draw a reference vector, representing a factor, which best summarizes the two vectors shown in Fig. 1.7. Intuitively, it seems likely that the best reference vector we can draw to summarize the two vectors in Fig. 1.7 would be the reference vector that bisects the angle, 25°, between the two vectors. This gives rise to the representation shown in Fig. 3.1 (with only half the vectors shown).

In Fig. 3.1, the variable 'temperature' is represented by T; 'hours of sunshine' by S and the 'reference vector' or factor by F_1. It can be seen from Fig. 3.1, that F_1 makes an angle of 12½° with T and with S. The cosine of 12½°, therefore, expresses the correlation between T and F_1 and between S and F_1. From Table 1.5, it will be seen that the cosine of 12½° = 0.976. In other words, the correlation between F_1 and T and F_1 and S is equal to 0.976, so it would seem that F_1 is a very good description of both T and S.

In the language of factor analysis, the correlation between a variable and a factor is known as the *loading* of the variable on the factor. Thus, the loading of S on F_1 is 0.976, which is the same as the loading of T on F_1. It was also noted in Chapter 1 that the 'square of the correlation coefficient', R^2, expresses the amount of variance

$$\hat{a} = \hat{b} = 12\tfrac{1}{2}°$$

Figure 3.1 *Vector diagram showing first reference vector, F_1, drawn to bisect the angle between T and S.*

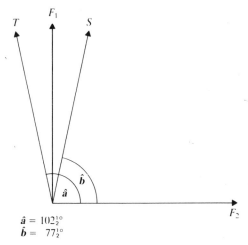

Figure 3.2 *Vector diagram showing two reference vectors, F_1 and F_2, required to resolve T and S.*

shared by two variables. Thus, the amount of variance shared by T and F_1 is $(0.976)^2 \simeq 0.95$. This is known as the *common factor variance* between T and F_1. Similarly, the common factor variance between S and F_1 is approximately equal to 0.95. The sums of the squares of the loadings of T and S on $F_1 = 0.976^2 + 0.976^2 = 1.9$. This is known as the *extracted variance* and it is the amount of common variance extracted by F_1 across T and S. If T and S each has a total variance of '1', the maximum variance that could be extracted by F_1 is equal to $1 + 1$, that is, 2. The percentage of variance extracted by F_1 is therefore equal to:

$$\frac{1.9}{2} \times 100 = 95\%$$

It would seem that F_1 is a very good summary of T and S. It gives us 95 per cent of the picture of the relationship between them. To get the complete picture, we would have to draw another vector, F_2, at right angles (or orthogonal) to F_1. In practical terms this wouldn't be very useful since all we would have succeeded in doing is replacing T and S by two reference vectors or factors, F_1 and F_2. The idea of factor analysis is to reduce the number of variables, not to replace all the original ones. Nevertheless, for completeness, F_2 has been drawn in Fig. 3.2.

The angle between T and $F_2 = 90° + 12\frac{1}{2}° = 102\frac{1}{2}°$ and the angle

Table 3.1
Principal components analysis for 'temperature', T, and
'hours of sunshine', S

Variables	Factors		Communality
	1	2	
T	0.976	−0.216	1.0
S	0.976	0.216	1.0
Extracted variance	1.9	0.1	2.0
Percentage of variance	95	5	100

between S and $F_2 = 90° − 12\frac{1}{2}° = 77\frac{1}{2}°$. Thus, the loading of T on $F_2 = \cos 102\frac{1}{2} = −\cos(180° − 102\frac{1}{2}°) = −0.216$. The loading of S on $F_2 = \cos 77\frac{1}{2}° = 0.216$.

Thus the amount of variance extracted by $F_2 = −0.216^2 + 0.216^2 = 0.1$, and the percentage of variance extracted by $F_2 = 0.1/2 \times 100 = 5$ per cent. In other words, F_2 has extracted the variance which remained unextracted by F_1.

In effect, what we have carried out is a *principal components analysis* of the variables T and S. The results of this analysis are summarized in Table 3.1. Only the final column, known as the *communality*, needs to be explained.

The communality of T is found by the sum of its common factor variance, that is, $0.976^2 + (−0.216)^2$, which equals 1. Similarly, the communality of S equals 1. The communality of a variable is the amount of variance it shares with all other variables and the maximum value it can reach is 1. It is hardly surprising that the communality of T and that of S is equal to 1 since we know a priori that T and S share *all* the common variance we attempted to extract by F_1 and F_2.

Let's now turn to how principal components can be used to reduce a 'large' number of variables to a smaller number of factors without losing too much information.

For example, in the personal alarms study, respondents were invited to 'rate' eight personal alarms on the following 15 statements, by simply stating whether they thought each of the statements applied to each of the alarms. The reader will notice that the statements are in a different order. This makes no material difference, but simply reflects the order in which they were put into the computer.

1. Feels comfortable in the hand
2. Could be easily kept in the pocket
3. Would fit easily into a handbag
4. Could be easily worn on the person
5. Could be carried to be very handy when needed
6. Could be set off almost as a reflex action
7. Would be difficult for an attacker to take it off me
8. I could keep a very firm grip on it if attacked
9. An attacker might be frightened that I might attack him with it
10. Would be difficult for an attacker to switch off
11. Solidly built
12. Would be difficult to break
13. Looks as if it would give off a very loud noise
14. An attacker might have second thoughts about attacking me if he saw me with it
15. I would be embarrassed to carry it round with me

The question is 'can these 15 variables be reduced to a smaller number of derived variables, known as factors, in such a way that we don't lose too much information?' The starting point is to look at how these variables correlate with each other. So far, we've only looked at the correlation between two variables but the basic idea is easily extended to the intercorrelations between more than two variables. For example, with three variables, 1, 2 and 3, we would need to compute the correlations between 1 and 2, 1 and 3, and 2 and 3. In other words we would take the variables in pairs, making sure we cover all possible pairs, and compute the correlations between them.

In our case of 15 variables, we have to compute $(15 \times 14)/2 = 105$ correlations to cover all possible pairs. In the survey, each of the eight personal alarms was rated by 100 respondents, so in total we have $8 \times 100 = 800$ observations from which the correlations are computed.*

The intercorrelations between a set of variables are usually presented in what is known as a *correlation matrix*. The correlation

---

*In each observation, each of the variables received either a score of '1', when a respondent thought it applied to a model, or '0' when a respondent thought it did not apply to a model. The coefficient computed to estimate the correlation between variables scored in this way is known as the four-fold point correlation. It is simply the product moment correlation between two variables for which the only admissible scores are 0 and 1.

matrix obtained for the 15 variables is shown in Table 3.2. The first thing to notice in Table 3.2 is that all the correlations could have been repeated in the top half since they are symmetrical about the diagonal which leads from 'top left' to 'bottom right'. In fact, this diagonal is called the leading diagonal, and it is usually included to reinforce the point that a variable correlated with itself gives a perfect correlation of 1.0.

To attempt an understanding of the complex interrelationships between the 15 variables simply by inspection is a pretty tedious and imprecise exercise, and, in any case, this is exactly what principal components analysis and factor analysis are designed to do for us! Nevertheless, some useful insights can be gained by looking at the 'high' correlations in the matrix.

One possible way is to look for blocks of 'significant' correlations. As we saw in Chapter 1, significance has a particular and precise meaning to statisticians and it relates to the notion of chance. In our case we would want to know how much greater or less than '0' a correlation has to be for us to be confident that it expresses a significant relation between two variables rather than one that could have occurred by chance.

Not surprisingly, statisticians have developed formulae to calculate this figure precisely for different sample sizes. These needn't concern us here but the appropriate figure as it relates to the data in Table 3.2 is that any correlation between two variables above 0.1 (or below −0.1) is statistically significant at the 95 per cent confidence level. Inspection of Table 3.2 also indicates that this figure wouldn't help us pick out the 'high' correlations given the overall level of the correlations in the matrix, since approximately three-quarters of the correlations are greater than 0.1 or less than −0.1. Another possibility is for us to take the top 25 per cent (the *upper quartile* as it is known). In our case this is (approximately) any correlation above 0.35 or below −0.35. In Table 3.3, correlations above 0.35 or below −0.35 shown in Table 3.2 are printed in bold.

Firstly, looking at the highest correlations (over 0.6 or below −0.6) we can see that variables 1 and 5 are highly correlated. In other words, when respondents rated an alarm as 'feels comfortable in the hand', they were also likely to rate it as 'could be carried to be very handy when needed'; so we might be tempted to conclude that both these variables are related to a notion of 'hand-feel', for want of a better description. We can also see that variables 2 and 3 are highly correlated. In other words, 'could easily be kept in the

Table 3.2
Correlation matrix between 15 statements describing personal alarms

	1	2	3	4	5	6	7	8	9	10	11	12	13	14	15
1	1														
2	.469	1													
3	.52	.686	1												
4	.324	.211	.29	1											
5	.666	.552	.545	.325	1										
6	.389	.331	.375	.29	.376	1									
7	.39	.234	.274	.31	.387	.317	1								
8	.509	.279	.315	.323	.48	.37	.612	1							
9	−.223	−.357	−.347	.001	−.221	−.109	.021	−.009	1						
10	.003	−.125	.015	.358	−.016	.156	.109	.126	.294	1					
11	−.212	−.232	−.207	−.021	−.184	−.03	.01	.01	.337	.289	1				
12	−.293	−.303	−.295	−.059	−.247	−.083	−.013	−.042	.4	.252	.651	1			
13	−.093	−.168	−.117	.112	−.063	.039	.003	.042	.223	.205	.178	.178	1		
14	.017	−.065	−.046	.156	.026	.133	.197	.161	.372	.234	.183	.177	.209	1	
15	−.401	−.488	−.448	−.154	−.382	−.226	−.145	−.216	.305	.097	.216	.275	.072	.055	1

Table 3.3
Correlations above 0.35 or below −0.35 (from Table 3.2), in bold type

	1	2	3	4	5	6	7	8	9	10	11	12	13	14	15
1	1														
2	**.469**	1													
3	**.52**	**.686**	1												
4	.324	.211	.29	1											
5	**.666**	**.552**	**.545**	.325	1										
6	**.389**	.331	**.375**	.29	**.376**	1									
7	**.391**	.234	.274	.31	**.387**	.317	1								
8	**.509**	.279	.315	.323	**.48**	.37	**.612**	1							
9	−.233	**−.357**	−.347	.001	−.221	−.109	.021	−.009	1						
10	.003	−.125	.015	**.358**	−.016	.156	.109	.126	.294	1					
11	−.212	−.232	−.207	−.021	−.184	−.03	.01	.01	.337	.289	1				
12	−.293	−.303	−.295	−.059	−.247	−.083	−.013	−.042	**.4**	.252	**.651**	1			
13	−.093	−.168	−.117	.112	−.063	.039	.003	.042	.223	.205	.178	.178	1		
14	.017	−.065	−.046	.156	.026	.133	.197	.161	**.372**	.234	.183	.177	.209	1	
15	**−.401**	**−.488**	**−.448**	−.154	**−.382**	−.226	−.145	−.216	.305	.097	.216	.275	.072	.055	1

pocket' and 'would fit easily into a handbag' tend to go together, perhaps suggesting something about 'size'. Similarly, 7 and 8— 'would be difficult for an attacker to take it off me' and 'I could keep a very firm grip of it if attacked'—are also highly correlated. Maybe together these suggest something about 'ease of holding in the hand'. Variables 11 and 12 also tend to go hand in hand—'solidly built' and 'would be difficult to break'. They seem to suggest 'robustness'.

Having examined the correlations above 0.6 (or below −0.6) we could then begin to examine those between 0.5 and 0.6 (or between −0.5 and −0.6), to look for a 'second-order level of togetherness', as it were. When we do this, the whole process becomes rather more complicated because we get certain relationships at this level that overlap relationships at the more stringent level of above 0.6 (or below −0.6). When we go down to correlations between 0.4 and 0.5 (or between −0.4 and −0.5), the picture becomes even more fuzzy. The idea of principal components analysis and factor analysis is to look for groups of variables in a more efficient and precise way than can ever be achieved by 'eyeball'. How is this done? Again, representing variables by vectors with the correlations being represented by the cosines of the angles between them is most useful.

Table 3.4 is a transformation of Table 3.2, where the correlations between the 15 variables are shown to the nearest 'whole' angle. Notice first of all that the leading diagonal now consists of a line of zeros, since two vectors which are correlated lie on top of each other.

Using Table 3.4, it is a relatively easy matter to draw as vectors the correlation between any two variables. For example, the correlation between 1 and 2 can be represented in two dimensions by two vectors at an angle of 62°. With three variables, for example 1, 2 and 3, things get a little more complicated, but we could, if we wanted, build a model in three dimensions, using three matchsticks and glue, in which the correlation between variables 1 and 2 is represented by two vectors at 62° to each other, the correlation between variables 1 and 3 by two vectors at 58° to each other and the correlation between 2 and 3 by two vectors at 47° to each other. With more than three vectors, the model becomes increasingly complex, since it is impossible to draw or construct a model in more than three dimensions. It requires what mathematicians call *multidimensional* space or *hyperspace* to describe a model which takes account of all the angular combinations shown in Table 3.4. Even

Table 3.4
Correlation coefficients represented as angles between vectors (from Table 3.2)

	1	2	3	4	5	6	7	8	9	10	11	12	13	14	15
1	0														
2	62	0													
3	58	47	0												
4	70	77	73	0											
5	48	56	57	71	0										
6	67	70	68	73	68	0									
7	67	76	74	72	67	73	0								
8	60	74	72	71	61	68	52	0							
9	103	111	110	90	103	96	89	90	0						
10	89	97	89	69	91	81	84	83	73	0					
11	103	103	102	91	101	92	89	89	70	73	0				
12	107	108	107	93	104	95	91	93	66	75	49	0			
13	95	100	97	84	94	88	90	87	77	78	80	80	0		
14	89	94	93	81	89	82	78	81	68	76	79	79	78	0	
15	124	119	117	99	122	103	98	103	72	84	77	74	86	87	0

though *hyperspace* models cannot be represented or thought of visually, they are a mathematical possibility. Though the mathematics will not concern us, it is important to grasp intuitively that hyperspace models can be constructed mathematically.

Imagine that all the intercorrelations between the 15 variables are represented by a hyperspace model, specifying all the angles between 15 vectors. The question that principal components analysis and factor analysis address is, 'how can these 15 vectors be replaced by a smaller number of vectors acting as reference vectors or factors to all 15 vectors?'

Earlier in the chapter, taking 'temperature' and 'hours of sunshine' as our example, we saw how factors might be constructed in two dimensions. A useful physical analogy to help get an idea of how vectors are resolved into factors in several dimensions is to think of a half-open umbrella with the radiating spokes hanging down at different angles to the central pole or handle. In this analogy, the radiating spokes of the umbrella represent vectors and the handle represents a resolution of these vectors, a reference vector, known as a factor. One important aspect of this analogy is to realize that the position of each of the spokes (vectors) is specified in relation to the handle (factor) by the angle between each spoke and the handle. As we saw earlier, in the language of principal components analysis and factor analysis, the cosine of the angle between a spoke (a vector) and the handle (a factor) is known as the *loading* of the vector on the factor, and thus *the loading of a variable on a factor expresses the correlation between the variable and the factor.* This is a very important thing to remember since, as we shall see, the loadings of variables on factors is critical to an understanding of the meaning and significance of factors.

To continue with our analogy, we can also see that some of the spokes (vectors) will load more highly on the handle (factor) than others. Spokes (vectors) which make small acute angles with the handle (factor) will load more highly than spokes (vectors) which make larger acute angles. In other words, our factor, represented by the handle, will give a better description of some vectors (represented by spokes making small acute angles with the handle) than others (represented by spokes making larger acute angles with the handle). To get a better resolution of the spokes (vectors) making larger acute angles with the handle, we would need a second handle (or factor), so to speak. Of course, while this second handle would resolve some spokes (vectors) better than the first one, by the same token, it would resolve others less well. From the comments so far in this paragraph, it is reasonably clear that an almost infinite number of handles or reference vectors could be used to resolve each of the spokes in different ways. Any one of these handles or reference vectors would resolve some spokes, or vectors, better than others. One of the questions that arises, therefore, is which handles or factors is it optimum to take? This is a question we will address in a few paragraphs' time.

Like all analogies, the half-open umbrella is fine as far as it goes. We've already stretched the idea of a half-open umbrella to a 'mythological' umbrella with several handles, and for the analogy to describe adequately the angular combinations between the vectors in Table 3.4, we would have to stretch it even further. For example, in Table 3.4 we can see that some vectors are at acute angles to each other (like the spokes of a real half-open umbrella), but others are at obtuse angles to each other (i.e. negatively correlated.) To take account of negatively correlated variables (vectors at obtuse angles to each other), we would have to imagine a higgledy-piggledy half-open umbrella with spokes at all different angles and with several handles—a sort of series of Siamese half-open umbrellas, each member being joined at its tip to every other member. The analogy is now becoming very bizarre indeed! And it doesn't stop here. Our series of Siamese half-open umbrellas can only be taken to represent vectors in three dimensions. To account for hyperspace, the mind begins to boggle. Nevertheless, the analogy is useful if it gets across the following three important ideas:

1. A large number of vectors (variables) can be resolved into a smaller number of reference vectors known as factors
2. The loadings of variables on factors express the correlations between the variables on the factors
3. Different variables will load differently on different factors

Once these three basic ideas have been taken on board, the next questions we need to ask ourselves are: which factors should be extracted? (bear in mind, as we noted two paragraphs back, there is nothing to stop us extracting any number of factors, each of which resolves the different variables more or less well), and just how many factors should we extract to arrive at an optimum solution? We will attempt to answer the first of these questions now, but the second question will be answered later.

It was mentioned a couple of pages back that the loading of a variable on a factor expresses the correlation between the variable and the factor. It is this finding that gives the key to the question of which factors should be extracted. We've also noted that the 'square of the correlation' between two variables, R^2, expresses the amount of variability or variance accounted for by the correlation between them, i.e. the common factor variance. Accordingly, *the square of the loading of a variable on a factor* is an expression of the common factor variance between a variable and a factor. In this sense, the higher the loading a variable has on a factor, the better the description of the variable by the factor.

The idea behind principal components analysis is to extract factors, sequentially, such that the first factor accounts for the maximum common factor variance across all the variables (i.e. more than any other factor that could be extracted). In principal components analysis a second factor (reference vector) is then extracted which is at right angles to the first factor (i.e. orthogonal to it) such that it extracts the maximum amount of the common factor variance that remains. A third factor is then constructed, again at right angles, and so on, until all the common factor variance has been extracted (or enough as makes no difference).

As we mentioned earlier, there is nothing sacred about the position of factors (reference vectors) and any factor in any position relative to the first factor will give a set of loadings of the variables which could be used just as well as the loadings on the orthogonal factors. There are, however, convenient mathematical properties associated with orthogonal factors and, while they needn't concern us, these properties explain why principal components analysis extracts orthogonal factors.

There's no doubt that the past few pages have been a trifle abstract so let's try to make some of these ideas more concrete by looking at a principal components analysis of the correlation matrix shown in Table 3.2.

A principal components analysis of Table 3.2

The correlation matrix (Table 3.2) is the starting point for a principal components analysis. The principal components matrix in which the first four principal components have been extracted is shown in Table 3.5, pretty much as it might come off a computer. (Note the similarity in format between Tables 3.5 and 3.1.)

To recap, it has been mentioned that the square of the loading of a

Table 3.5
Principal components matrix for the correlations in Table 3.2

	Factors				
Variables	1	2	3	4	Communality
1	.785	.118	.024	.115	.644
2	.757	−.135	.186	−.231	.68
3	.776	−.032	.103	−.293	.699
4	.431	.442	−.352	−.237	.562
5	.787	.126	.081	.051	.645
6	.541	.316	.001	−.121	.407
7	.503	.458	.118	.471	.699
8	.585	.46	.118	.385	.716
9	−.438	.567	−.098	.187	.558
10	−.053	.633	−.222	−.396	.61
11	−.377	.564	.545	−.242	.816
12	−.471	.54	.51	−.162	.799
13	−.161	.415	−.406	−.224	.413
14	−.033	.58	−.253	.191	.438
15	−.606	.151	−.06	.263	.462
Latent root*	4.438	2.628	1.053	1.029	9.148
Percentage variance	29.6	17.5	7	6.9	61

*Also known as extracted variance, eigenvalue or sum of squares.

variable on a factor is an expression of common factor variance between a variable and a factor. Thus, the sum of the squares of all the loadings of all the variables on a factor gives us the amount of common variance extracted by a factor across all the variables. For example, the sum of squares of the loadings, the common factor variance for factor 1, is calculated as follows:

$$(0.785)^2 + (0.757)^2 + (0.776)^2 + (0.431)^2 + (0.787)^2 + (0.541)^2 + (0.503)^2 + (0.585)^2 + (-0.438)^2 + (-0.053)^2 + (-0.377)^2 + (-0.471)^2 + (-0.161)^2 + (-0.033)^2 + (-0.606)^2$$

If you can be bothered, you'll find that this sum comes to 4.438, as shown in the penultimate row at the bottom of the factor 1 column in Table 3.5. You will also see from Table 3.5 that this sum is referred to as the *latent root* or, alternatively, *eigenvalue, extracted variance* or *sum of squares.* (Latent root and eigenvalue are expressions taken from matrix algebra and they won't concern us.)

Firstly, note that the values of the sum of squares of the loadings fall off from the first factor. This is because, as we've already noted, in principal components analysis the maximum amount of variance is extracted by each factor in turn, starting with the first factor.

For 15 variables, each with a total variance of '1', the maximum possible variance that could be extracted by any one factor is 15. The actual variance extracted by the first factor is 4.438.

A common way of expressing the latent roots (sum of squares of the loadings) is to convert them to percentages of the maximum possible variance. For the first factor, the percentage variance is:

$$\frac{\text{Latent root}}{\text{Number of variables}} \times 100 = \frac{4.438}{15} \times 100 = 29.6\%$$

The percentage variances calculated for each factor are shown in the final row of Table 3.5. Percentage variance is a useful statistic because it gives a clear idea of the contribution of the factors to the total variance. For example, the first four factors in our analysis account for:

$$29.6\% + 17.5\% + 7\% + 6.9\% = 61\%$$

Now, let's turn to the final column in Table 3.5, headed 'communality'. As we noted earlier, the communality of a variable is, in fact, the sum of all the common factor variance of the variable. So, for example, the communality of variable 1, shown as 0.644 in Table 3.5, is equal to the sum of:

$$(0.785)^2 + (0.118)^2 + (0.024)^2 + (0.115)^2$$

In other words, the communality of a variable is the variance it shares in common with the other variables. Accordingly, 64.4 per cent of the variance of variable 1 is shared with the other 14 variables. As we have noted on a few occasions already in this chapter, the aim of factor analysis is to discover common factors. Mathematically, therefore, if the communality of a variable is too low, say 0.3 or less, we might feel it does not contribute enough to warrant inclusion in a factor analysis. In our example, none of the communalities of the variables falls to such a low level. From now on, we shall not have occasion to refer to communality.

Before we begin to examine the main body of Table 3.5 in any detail, that is, the factor loadings for each factor and each variable, we'll pick up the question we left hanging in the air earlier, namely, how many factors should we extract? In our case we have extracted four factors which together account for approximately 60 per cent of the variance; but why did we stop there?

The answer to this question has given rise to a fair bit of controversy among the experts and some of the mathematical arguments are difficult to follow. It is also true to say that there isn't a single definitive answer to the question—you pay your money and take your choice!

One thing everyone would agree on, however, is that the factor solution you decide to 'run with' must explain at least the majority of the variance. Just how much is, again, a matter of opinion and in the final analysis this criterion is rather arbitrary. In market research, it is not uncommon to find factor solutions which account for no more than 50 per cent of the variance, and sometimes less than this. Some people would consider this less than satisfactory and therefore, they apply an arbitrary 75 per cent rule.

A rule that is very simple to apply and one that is popular among some factor analysts is to extract only *factors having latent roots greater than one*. In our example, only the first four factors satisfy this criterion. It has been suggested that the criterion is most reliable when the number of variables is between 20 and 50. Where the number is less than 20, as in our case, there is a not too serious tendency to extract a conservative number of factors; conversely, where the number of variables is greater than 50, the tendency is to extract too many factors.

Irrespective of which criteria are applied—and there are many of them—ultimately the factor solution has to make sense for it to be of any use. Indeed, in market research, the criterion 'can I make sense of it' often overrules any stricter mathematical criteria, to the extent that a cynic might be tempted to agree with certain critics, who believe that factor analysis as practised by some researchers is a wholly subjective exercise with just a cloak of respectability provided by the computational sophistication used to derive correlations and factor loadings. Clearly, 'making sense' is a necessary condition but it is hardly sufficient on its own. My advice to any recipient of a study based on factor analysis is to ask two fundamental questions: what is the amount of variance accounted for; and what *objective* criterion was used to determine the factor solution presented? Failure to get satisfactory answers to these two questions might leave some doubts!

Interpreting the factor loadings

Since the loading of a variable on a factor represents the correlations between the variable and the factor concerned, within any one factor, we're obviously interested in those variables with high loadings. Before we examine the size of the factor loadings in Table 3.5 in any detail, it is worth noting that in all four factors there are both positive and negative loadings (i.e. positive and negative correlations). Factors with both positive and negative loadings are known as *bipolar factors* because the factors contain contrasting or opposite groups of variables. In terms of the geometry we used earlier, some of the vectors have been resolved in one direction and others in opposite quadrants, giving rise to positive and negative values. There is nothing absolute in these signs, so provided we are consistent within a factor, we could, if we wanted, make all the positive signs negative and vice versa. For example, in factor 1, we could reverse the signs of variables 1 to 8 and make them negative, provided we also reverse the signs of variables 9 to 15 and make them positive. (Reversals of this sort are sometimes done to make the largest loadings positive.) Of course, not all factor analyses produce bipolar factors; this just happens to be the case in our example. Sometimes all the signs of the loadings in a factor are positive and sometimes they are all negative. Such factors are known as *unipolar* factors. In the latter case this does not mean we've found a 'negative factor'. All that has happened is that the computer has 'read the angles' between the variable vectors and the factor vectors with the factor vector at 180° to the usual position. In these cases, it is usual practice to reverse all the negative signs, that is, to ignore them. It is only in factors where positive and negative loadings are found together (bipolar factors) that the relativities of the different signs have any significance.

Now let's 'try to make sense' of the factor loadings shown in Table 3.5. Since factor loadings represent correlations, the first question we want to ask is which of the loadings are statistically significant? Once again, not surprisingly, statisticians have produced formulae for this purpose, but the issues are not quite as straightforward as those involved in calculating the significance of a correlation between two variables. Rather than getting side-tracked into these issues, fortunately, there's a 'rule of thumb' which is usually acceptable for the data generated in survey research studies. This rule states that only loadings above 0.3 or below −0.3 should be considered as significant. This is a fairly stringent rule for sample sizes greater than 50, but in the 'rough and ready' world in which survey research data are often collected, it does no harm to apply stringent rules.

Table 3.6 shows the same figures as those in Table 3.5 only this time the significant loadings are shown in bold type. In the first factor all but the following variables load significantly:

10. Would be difficult for an attacker to switch off
13. Looks as if it would give off a loud noise
14. An attacker might have second thoughts about attacking me if he saw me with it

To be honest, eliminating these variables doesn't make the 'meaning' of this factor very obvious. Sometimes a useful way of trying to get to grips with the 'meaning' of a factor (if indeed there is one to get to grips with!) is to list the variables in descending order of their loadings. This is done below, leaving out the variables that do not load significantly.

5. Could be carried to be very handy when needed
1. Feels comfortable in the hand
3. Would fit easily into a handbag
2. Could be easily kept in the pocket
8. I could keep a very firm grip on it if attacked
6. Could be set off almost as a reflex action
7. Would be difficult for an attacker to take it off me
4. Could be easily worn on the person
11. Solidly built
9. An attacker might be frightened that I might attack him with it
12. Would be difficult to break
15. I would be embarrassed to carry it round with me

If we remember that this factor is a bipolar factor and, moreover, the last four variables load negatively, we might begin to speculate that we have identified a factor of 'size' which spans a continuum from 'small' to 'large'. However it also has to be said that if such a

Table 3.6
Principal components matrix for the correlations in Table 3.2 with significant loadings in bold type

Variables	Factors			
	1	2	3	4
1	**.785**	.118	.024	.115
2	**.757**	−.135	.186	−.231
3	**.776**	−.032	.103	−.293
4	**.431**	**.442**	**−.352**	−.237
5	**.787**	.126	.081	.051
6	**.541**	**.316**	.001	−.121
7	**.503**	**.458**	.118	**.471**
8	**.585**	**.46**	.118	**.385**
9	**−.438**	**.567**	−.098	.187
10	−.053	**.633**	−.222	**−.396**
11	**−.377**	**.564**	**.545**	−.242
12	**−.471**	**.54**	**.51**	−.162
13	−.161	**.415**	**−.406**	−.224
14	−.033	**.58**	−.253	.191
15	**−.606**	.151	−.06	.263
Latent root*	4.438	2.628	1.053	1.029
Percentage variance	29.6	17.5	7	6.9

* Also known as extracted variance, eigenvalue or sum of squares.

factor 'exists', on which people judge personal alarms, then the 15 variables on which our sample was asked to rate the alarms is not a particularly good measure of 'size', since we might think that variables 10, 13 and 14 relate to size as well. Certainly, we would be very cavalier indeed to conclude that we have identified a general factor of size that applies to all personal alarms.

If we return to the correlation matrix in Table 3.2 we can see that there are some close similarities between the large correlations and the first factor. At least superficially, it seems that the first factor represents the general interrelationships between the variables.

Turning now to the second factor we see first of all that this is actually a unipolar factor and not a bipolar one, since none of the negative loadings is significant once we've applied our rule of thumb about the significance of factor loadings. Just as we did with the loadings in factor 1, we can list the variables in factor 2 in descending order of their loadings. This is done below.

10. Would be difficult for an attacker to switch off
14. An attacker might have sound thoughts about attacking me if he saw me with it
 9. An attacker might be frightened that I might attack him with it
11. Solidly built
12. Would be difficult to break
 8. I could keep a very firm grip of it if attacked
 7. Would be difficult for an attacker to take it off me
 4. Could be easily worn on the person
13. Looks as if it would give off a very loud noise
 6. Could be set off almost as a reflex action

The variables with high loadings on this factor correspond to those variables with negative loadings on factor 1, i.e. those variables we hypothesized were measuring 'largeness' on a 'small–large' continuum. Moreover, we might also note that the variables with high positive loadings on factor 1 (i.e. those we thought might be measuring 'smallness') do not load significantly on factor 2.

Considering these two observations together, we might begin to speculate that factor 2 is measuring something related to 'largeness'. Looking at the 'meaning' of the variables which load most highly on factor 2, we could begin to speculate further, as researchers are wont to do, that factor 2 is not just concerned with size but also with the 'appearance' of personal alarms, in some way pertaining to how 'offensive' a personal alarm is perceived to be. (Many people might reasonably think that this line of argument is somewhat disingenuous!)

To complete our inspection of Table 3.6, we will again list the significant loadings on factor 3 and factor 4 (which are bipolar factors) in descending order.

Factor 3
11. Solidly built
12. Would be difficult to break

4. Could be easily worn on the person
13. Looks as if it would give off a very loud noise

Factor 4
7. Would be difficult for an attacker to take it off me
8. I could keep a very firm grip of it if attacked
10. Would be difficult for an attacker to switch off

The 'meanings' of these factors appear to add very little to what we have learned from factors 1 and 2.

Before leaving this section, it is perhaps worth re-emphasizing that we have not found any *objective* evidence to support the view that people judge personal alarms on four clear factors or dimensions. Of course, factor analysis can only work on the variables entered into the analysis, and if these are insufficient or inadequate in spanning the range of dimensions or attributes on which people judge the things in question (in our case personal alarms), then this is a failing of the researcher who 'chose' the variables, and not of factor analysis. In our example, based on the results of the factor analysis so far presented, if we were to repeat the study, we might well wish to include additional statements which we felt tapped into perceptions of 'size' and 'offensiveness' in the hope that a subsequent factor analysis would identify them as two clear factors.

Rotation of factors

Principal components analysis is sometimes referred to as a *direct* method of extracting factors because the factor matrix thus obtained (e.g. Table 3.5) is as a direct result of applying a particular mathematical procedure to extract orthogonal factors from the correlation matrix (e.g. Table 3.2). In the previous section it was mentioned that there is nothing sacred about orthogonal factors and, moreover, that any number of different factors (i.e. reference vectors in different positions) will give a set of loadings which might be used just as well as the loadings on orthogonal factors. An obvious question that arises, therefore, is whether the orthogonal factors produced by principal components analysis give the best picture of the interrelationships between a set of variables?

The majority of factor analysts agree that in certain situations direct methods do not provide the most illuminating picture concerning the interrelationships between a set of variables. In our own example, considered in the previous section, our interpretation of the factor matrix in Table 3.5 was far from clear-cut, primarily because several variables loaded significantly on more than one factor. In such a situation it usually pays to look at rearrangements of the factors (i.e. rearrangements of the reference vectors) to see if we can reduce some of the ambiguities. The process whereby the factors (reference vectors) are rearranged is known as *rotation*. The results of rotation are sometimes known as *derived* solutions, which reflects the fact that they are derived from direct solutions.

'Rotation' means exactly what it says, that is, the factors (reference vectors) are turned about their origin through hyperspace, into a different position. The simplest form of rotation is when the factors (reference vectors) are kept at 90° to each other and are moved jointly, giving rise to what is known as *orthogonal rotation*.

It is also possible, and increasingly popular, to rotate the factors through different oblique angles. Not surprisingly, this is known as *oblique rotation*.

Typically, having carried out a principal components analysis and having inspected the resultant factor matrix only to find an ambiguous picture (as we have), the next step is to try an orthogonal rotation. We'll do this shortly, but before proceeding, the remarks in the previous paragraph beg a rather important question, namely, during rotation how do you know when you've arrived at an alternative position that is optimum? For two factors this is achieved in effect, by maximizing the 'sums of the squares of the loadings' of some variables on one factor and, simultaneously minimizing the 'sums of the squares of the loadings' of these variables on the second factor, and vice versa. With more than two factors the situation is more complicated. Essentially, all the factors are taken together in pairs and rotated until the optimum position is reached such that the loadings of all the variables on all the factors is 'least ambiguous'. Not surprisingly, there are different ways of carrying out this process of optimization, each with its pros and cons. A common orthogonal method of rotation is known as the *Varimax* procedure and the results of a Varimax analysis for the data in Table 3.2 are given in Table 3.7. Again, applying our rule of thumb for the significance of factor loadings, the loadings above 0.3 and below −0.3 are in bold type.

A comparison of Tables 3.6 and 3.7 reveals that the overall percentage of variance accounted for remains the same between the principal components solution and the Varimax solution (61 per cent). This is to be expected, of course, since altering the position of

Table 3.7
Varimax solution for correlations in Table 3.2 with significant loadings in bold type

Variables	Factors			
	1	2	3	4
1	**.56**	.018	−.2	**.538**
2	**.791**	−.114	−.104	.176
3	**.808**	.032	−.098	.187
4	**.345**	**.611**	−.093	.245
5	**.608**	.013	−.135	**.507**
6	**.463**	.258	.023	.354
7	.12	.042	.032	**.826**
8	.233	.077	.033	**.809**
9	**−.544**	**.363**	**.308**	.189
10	.047	**.719**	**.301**	.006
11	−.124	.127	**.886**	.000
12	−.247	.101	**.853**	−.01
13	−.147	**.621**	.034	−.066
14	−.255	**.468**	.072	**.385**
15	**−.661**	.032	.148	−.055
Percentage variance	21.8	11.7	12.1	15.4

the factors (reference vectors) does not alter the total variance. What it does alter, however, is the distribution of the variance across the four factors. For example, comparing the loadings of variable 1 in Tables 3.6 and 3.7, we can see that changing the position of the factors (reference vectors) has had the effect of transferring some of the variance from factors 1 and 2 to factors 3 and 4. As we would also have predicted, given the purpose of rotation, the loadings of moderate size in the principal components analysis have, by and large, increased or decreased to the point of insignificance, and there are fewer variables which load significantly on more than one factor. The ambiguities have not been resolved completely, however. For example, variable 1 loads significantly on factors 1 and 4; variable 4 loads significantly on factors 1 and 2; and an inspection of Table 3.7 reveals several other variables that load significantly on two or more factors. At this stage, rather than examining the structure of the loadings in any detail, it looks as if it would be worth trying another rotation, this time an oblique one. Before looking at an oblique solution, one or two preliminary comments will serve as an introduction.

There is still disagreement among factor analysts about how these oblique rotations should be carried out, and some factorists carry out more than one oblique analysis in arriving at a 'solution'. As the name 'oblique' implies, there is almost bound to be a correlation between the factors extracted. If this is the case, it is argued that the results of an initial oblique analysis (sometimes called a *primary* analysis) can themselves be treated as a correlation matrix for the purposes of a second factor extraction and oblique rotation. In theory, the process can be repeated a third time, and so on. Clearly, each additional factor extraction reduces the number of factors found in the previous extraction, until in the limited case only one factor is left. This often occurs at the fourth or fifth generation of extraction.

Finally, before looking at an oblique solution in any detail, the rationale that underpins an oblique rotation is that most, if not all, the variables being measured are correlated to some extent and, therefore, the underlying major factors must also be correlated.

For example, if our first tentative factor of 'size' and our second one of 'offensiveness' have any objective meaning (i.e. meaning independent of the factor analysis), then we would almost certainly expect them to be correlated to the extent that the larger a personal alarm is, the more likely it is to be perceived potentially as an offensive weapon.

The assumption of correlation between factors should not worry us, therefore, and some would argue the assumption is more 'realistic' than the assumption of 'orthogonality' inherent in principal components and orthogonal rotations.

Table 3.8 shows an oblique solution for our original correlations data shown in Table 3.2. In comparing Tables 3.7 and 3.8, it can be seen that while, again, the percentage variance extracted remains unaltered at 61 per cent, the distribution of the variance across the

Table 3.8
Oblique solution for correlations in Table 3.2 with significant loadings in
bold type

Variables	Factors			
	1	2	3	4
1	**.346**	.000	−.101	**.389**
2	**.676**	−.103	.033	−.001
3	**.694**	.04	.03	−.007
4	.258	**.601**	−.084	.101
5	**.412**	−.007	−.03	**.35**
6	**.355**	.233	.089	.215
7	−.099	−.029	.063	**.77**
8	.012	.01	.078	**.725**
9	**−.493**	**.308**	.192	.28
10	.12	**.689**	.244	.065
11	.06	.05	**.842**	.035
12	−.061	.026	**.793**	.055
13	−.098	**.616**	−.042	−.087
14	**−.316**	**.421**	−.000	**.389**
15	**−.578**	.01	.04	.097
Percentage variance	19.8	20.7	8.9	11.6

factors is different. The major difference between the Varimax
solution and the oblique solution is that the loadings in the oblique
solution have been pushed even higher or lower. (For example,
whereas variable 4 loads significantly on factors 1 and 2 in the
Varimax solution, in the oblique solution it loads significantly on
factor 2 only.) The overall effect of this is to reduce the number of
variables that load significantly on more than one factor. In the
Varimax solution, variables 1, 4, 5, 6, 9, 10 and 14 load significantly
on two or more factors. By comparison, in the oblique solution only
four variables—1, 5, 9 and 14—load significantly on more than one

factor. The oblique solution does look as if it has resolved many of the ambiguities found in the earlier solutions. Let's examine the oblique solution a little more closely to see if it 'makes sense'. As before, we'll list the variables in each of the four factors in descending order of their loadings.

Factor 1
3. Would fit easily into a handbag
2. Could be easily kept in the pocket
5. Could be carried to be very handy when needed
6. Could be set off almost as a reflex action
1. Feels comfortable in the hand
14. An attacker might have second thoughts about attacking me if he saw me with it
9. An attacker might be frightened that I might attack him with it
15. I would be embarrassed to carry it round with me.

Factor 2
10. Would be difficult for an attacker to switch off
13. Looks as if it would give off a very loud noise
4. Could be easily worn on the person
14. An attacker might have second thoughts about attacking me if he saw me with it
9. An attacker might be frightened that I might attack him with it

Factor 3
11. Solidly built
12. Would be difficult to break

Factor 4
7. Would be difficult for an attacker to take it off me
8. I could keep a very firm grip of it if attacked
14. An attacker might have second thoughts about attacking me if he saw me with it
1. Feels comfortable in the hand
5. Could be carried to be very handy when needed

As we noted before, factor 1 seems to be measuring a dimension of size, on a continuum from small to large. Factor 2 certainly appears to be tapping in to some aspect of the appearance of a personal alarm but our earlier speculation that this 'something' is to do with the 'offensive' appearance of a personal alarm seems no more apparent than it was when we examined the principal components analysis. Factor 3 now seems a lot clearer and we might reasonably

Table 3.9
Correlations between oblique factors

	1	2	3	4
1	1			
2	−.027	1		
3	−.382	.16	1	
4	.494	.144	−.174	1

decide to label this factor 'robustness'. Again, factor 4 appears far more coherent than it did when we examined the principal components analysis and looks as if it is connected with what we might call 'hand-feel'.

As we've already noted, in oblique solutions the factors are typically correlated and this must be remembered when interpreting oblique solutions. Table 3.9 shows the correlations between the oblique factors. Given what has already been said, it is hardly surprising that the 'size' of a personal alarm (factor 1) is correlated with its 'robustness' (factor 3) and its 'hand-feel' (factor 4).

Factor score weights

As we've seen repeatedly, the 15 variables load differently on each of the four factors, and these loadings vary according to the type of factor solution. For example, inspection of Table 3.8 indicates that variable 3 has approximately twice the loading of variable 1 on factor 1. However, this doesn't mean that variable 3 contributes approximately 'twice as much' to factor 1 compared with variable 1. Remember, factor loadings represent correlations between variables and factors, and the 'square of the loadings' expresses the amounts of variability (or variance) accounted for by the correlations between them. Accordingly, we can calculate from Table 3.8 that of the total amount of variability accounted for by factor 1, variable 1 accounts for $(0.346)^2 = 0.12$ of it and $(0.694)^2 = 0.48$ is accounted for by variable 3. So, whereas on factor 1 the loading of variable 3 is approximately twice the loading of variable 1, the contribution of variable 3 to the total variance accounted for by factor 1 is not twice that of variable 1.

It is possible to calculate the relative 'weight' that each of the variables contributes to each of the factors, and these are known as *factor score weights*. How this is done needn't concern us, but the

Table 3.10
Factor score weights for oblique solution

Variables	Factors 1	2	3	4
1	0.019	−0.041	−0.031	0.197
2	0.401	−0.053	0.179	−0.155
3	0.424	0.053	0.149	−0.201
4	0.131	0.443	−0.151	−0.144
5	0.097	−0.05	0.041	0.149
6	0.175	0.124	0.079	−0.002
7	−0.335	−0.187	0.002	0.591
8	−0.244	−0.148	0.029	0.518
9	−0.307	0.11	−0.022	0.245
10	0.233	0.467	0.071	−0.269
11	0.316	−0.137	0.668	−0.051
12	0.218	−0.153	0.609	0.005
13	0.009	0.467	−0.181	−0.193
14	−0.303	0.216	−0.169	0.26
15	−0.363	−0.043	−0.087	0.219

outcome is a table of factor score weights such as the one shown in Table 3.10, which shows the factor score weights calculated for the oblique solution. A comparison of Tables 3.8 and 3.10 confirms what we would expect, namely, that variables with high loadings on a factor also tend to have high factor score weights on the same factor.

Factor score weights are useful in a number of ways, not the least of which, in our example, is that they permit a comparison between each of the personal alarms in terms of their 'profiles' on each of the factors. For any model on any factor, in effect, this is achieved by multiplying the percentage of respondents who thought each of the variables applied to a personal alarm by the corresponding factor score weights of the variables, and summing the resulting 15 products to arrive at factor scores for each of the models on the four factors. The resulting factor scores for the eight models are shown in Table 3.11. To show how this table can be used, as an example let's look at the profiles of the eight personal alarms on factor 1. We had suggested that factor 1 was measuring a dimension of size, on a continuum from small to large. In terms of their profiles in Table 3.11, if factor 1 is 'in reality' measuring a dimension of size, we would expect the smaller personal alarms to have larger positive scores and the larger ones to have larger negative scores. In terms of their profiles on factor 1, we can list the models from the highest to lowest scores as follows:

A, F, C, H, D, B, G, E

The question we can now ask is whether there is any objective evidence to support our suggestion that factor 1 is measuring size. If it is, we would expect to obtain the above ordering of the alarms, or an approximation to it, if we asked a sample of people to rank the eight alarms from smallest to largest. This was not done in the present study, but in a study carried out in parallel to it a group of experts including crime prevention officers and staff of Age Concern made the following unprompted comments about the size of the alarms:

Model A
'A good size but perhaps too small'
Model B
'Too big'
Model C
'Small, light, nice to hold'
Model D
'Small, but not too small, light and easy to use'
Model E
'Far too big—can only just be held'
Model F
'Small but uncomfortable to hold'
Model G
'Easy to hold but too big'
Model H
'If it were slightly smaller and gave off a good noise it would be the best'

Table 3.11

Profiles of eight personal alarms on factor score weights

| Models | Factors | | | |
	1	2	3	4
A	97.6	50.8	4.9	−56.3
B	−10.2	107.3	−94.1	26.3
C	57.7	−124.8	7.6	28.4
D	14.6	−100.7	−98.5	18.7
E	−156.6	−19.0	−5.2	−37.5
F	83.3	12.3	90.3	−112.8
G	−107.4	19.7	35.4	58.3
H	21.0	54.5	59.6	75.0

These comments do appear to provide some independent evidence which suggests that factor 1 is indeed measuring a dimension of size. Moreover, of passing interest, by looking at the profiles of the alarms on factors 2, 3 and 4 in the light of the other comments made about them by the experts, there is also corroborative evidence for the factors of 'robustness' (factor 3) and 'hand-feel' (factor 4). Factor 2 is not quite as clear-cut—while the experts certainly made comments about the appearance of the alarms, it was not easy to interpret the profiles of the models on factor 2 in the light of their comments.

The limitations of factor analysis

The rationale of factor analysis, particularly as it is frequently employed in survey research, has been a subject of controversy for years. In its crudest form, it is used to describe complex matrices of correlations (such as the one in Table 3.2) by factors chosen for purely mathematical reasons. For example, if one group of variables is correlated highly with each other but not at all with a second group of variables, it is sometimes argued that the intercorrelations are due to a single underlying factor which the variables of the first group assessed with success but which is not involved in the variables of the second group. In this case, the argument is based purely on the correlations without any knowledge whatsoever of the nature or meaning of the variables entered in the factor analysis. Unfortunately, as in our example of the personal alarms, matrices of correlations do not always fit into such simple patterns and in these circumstances it is open to debate as to which of several possible descriptions in terms of factors is the best one. To continue with the argument, if the second group of variables correlate with those in the first group but not as highly as the variables in the first group do with each other, the question arises as to how best to describe such a situation. Should it be described by one general factor measured to some degree by all the variables or should it be described by several factors? Both approaches would be mathematically acceptable. The decision as to which variables are the best measures of a factor is also ambiguous to some extent. Decisions taken on purely mathematical grounds as, for example, in choosing to extract orthogonal factors may be open to question since, as we saw earlier in the chapter, on psychological as opposed to mathematical grounds, in certain circumstances we might well expect factors to be correlated with each other. The difficulty is a more sophisticated form of the old problem of interpreting correlations which was raised in Chapter 1. There is a correlation between wearing spectacles and bad eyesight but we do not decide on the underlying cause of the correlation mathematically. If we did, we might conclude that wearing spectacles *causes* bad eyesight, and, accordingly, we would recommend that the wearing of spectacles should be discouraged! Rather, we interpret the mathematics by our knowledge of other facts. By the same token, we should choose the factors underlying matrices of correlations for reasons 'outside' the matrices.

In the particular data subjected to factor analysis in this chapter, an attempt to arrive at an outside criterion for 'choosing' the factors was made by looking at the comments made by experts about the personal alarms, independent of the factors generated in the factor analysis. In doing this, there was some independent evidence for a factor of 'size' on which the personal alarms were evaluated. In fact, the early development of factor analysis was driven by psychological considerations rather than mathematical ones, particularly in the fields of human intelligence and personality. In survey research, however, factor solutions are often driven by the mathematics or by *post hoc* interpretations of the factors derived mathematically. The problem in relying on *post hoc* interpretations is that it is inherent in human nature to look for patterns in data, and, as we saw in Chapter 1, it is even possible to 'see' patterns in random data. The main point being made here is that factor analysis is *not* a mathematical procedure, pure and simple. Rather, its success depends also on an understanding of the phenomena being investigated. The mathematics alone cannot guarantee a 'correct' result. By the same token, *post hoc* interpretations of mathematically derived factors cannot guarantee an acceptable result either. In survey research, perhaps the best we can do is to insist that the factors derived satisfy some basic mathematical criteria. At least, in the author's view, the number of factors extracted should be defensible on mathematical grounds, and the factor solution should account for the majority of the variance. Both these criteria have been discussed in this chapter.

4. Mapping techniques

Introduction

As we've seen, to represent the relationship between two variables, it is a fairly easy matter to draw a scatterplot in two dimensions, just as we did for 'temperature' and 'hours of sunshine' in Chapter 1. Alternatively, this relationship can be represented by vectors in two dimensions, where the angle between the vectors expresses the correlation between the variables. With three variables, we can represent the interrelationships by building a three-dimensional model using matchsticks and glue. However, there is no *direct* way of showing the relationships between four or more variables. Nevertheless, with four or more variables, we could, if we wanted, examine scatterplots for all pairs of variables as a simple way of looking at the data. Unfortunately, in situations where we have several variables to examine, this is usually unsatisfactory because the number of plots is large and examining them all can be a very confusing job. (For example, with seven variables, we would need to examine 21 scatterplots!) Moreover, such plots can in fact be misleading since any structure that might be present in the original hyperspace is not necessarily reflected in what we might observe in the scatterplots of the variables taken in pairs.

Therefore, given that there is no entirely satisfactory way of directly representing the interrelationships between more than three variables, the question that mathematicians have asked themselves is whether it is possible to derive an *indirect* representation of the interrelationships between more than three variables. The answer is that mathematicians have developed a variety of techniques, and in this chapter we'll look at a family of techniques known as *mapping* or *ordination* methods. These methods set out to provide a representation of the original hyperspace in a reduced number of dimensions while retaining the original structure found in hyperspace as far as possible. Typically, these techniques are used to produce a two-dimensional map of the original hyperspace, hence their generic name 'mapping techniques'. It is also possible to represent the original hyperspace in three dimensions by making a three-dimensional model but this method is less common and, typically, less useful.

Measuring the 'relationships' between variables and objects

In arriving at an estimate of the relationship between 'temperature' and 'hours of sunshine', we calculated the correlation coefficient. This is an estimate of how much these two variables 'go together' or are 'associated' with each other. In a sense, therefore, the correlation coefficient is a measure of how *similar* 'temperature' and 'hours of sunshine' are to each other.

Superficially, the idea of two or more things being similar or dissimilar to each other is commonplace. 'Similarity' and 'dissimilarity' are concepts we use frequently to classify objects, events or, more generally, 'things', to allow us to make sense of them. Indeed, our ability to classify things is part of our fundamental psychology; and without this ability, our daily lives would be impossible.

Imagine if we had to treat every instance of something, or every object we encountered, as unique. Of course we don't treat everything we meet as unique, even if the more sensitive side of our soul acknowledges that everything has certain unique qualities. What we tend to do is to classify things as belonging to this or that general class of things. How we do this is far from simple and the process of classification has dominated psychology and many branches of science for centuries. For example, on a workmanlike level, we don't always group things on the number of characteristics they share in common; sometimes we give more weight to certain characteristics than others; sometimes we group things on the basis of characteristics they don't possess, and so on. Indeed, the concept of similarity raises fundamental philosophical and conceptual issues, such as how can we best organize the world into useful categories of things that are similar to each other.

All mapping techniques (and clustering techniques, which we will consider in the next chapter) work on a 'measure' of how similar things such as objects and variables are to each other. In other words, these techniques take as their starting points estimates of the similarities between each and every object or variable under investigation. For example, principal components analysis first of all calculates the intercorrelations of all members in a set of variables and then groups the variables into factors on the basis of these correlations. Certain mapping methods work on the judgements of similarity of objects made by respondents. In Chapter 2, we considered a matrix of 'judged similarity' for the eight personal alarms.

The correlation coefficient is an indirect estimate of similarity in comparison with the 'judged similarity' of the eight personal alarms, which is a direct estimate of similarity. Mapping techniques work on either direct or indirect estimates of similarity, and in this chapter

Table 4.1
Scores used to compute the
correlation between models C
and D ($r = 0.98$)

Variables	Models	
	C	D
1	78	81
2	1	2
3	93	94
4	14	52
5	95	97
6	96	100
7	41	46
8	59	67
9	35	68
10	16	21
11	98	97
12	6	6
13	0	0
14	19	19
15	1	1

we'll consider mapping applications to both direct and indirect estimates.

Although the correlation coefficient is sometimes used as an indirect measure of similarity, it does have one or two drawbacks. For example, from the data in the personal alarm study, it is possible to calculate the correlation coefficient between models C and D on the 15 variables on which they were both judged. This is done by computing r from the data in Table 4.1. For C and D, $r = 0.98$. Certainly, using the correlation coefficient as a measure of similarity, models C and D appear to be very similar indeed to each other. However, if we examine the 'profiles' of C and D on their scores on the 15 variables, we can begin to see how the correlation coefficient loses some information concerning the similarity between C and D. On the first ten variables, model D scores more highly than model C. In other words, on these ten variables, the 'elevation' of model D is higher than that of model C, and this is particularly so on variables 4 and 9. On variables 11 to 15 the profiles of models C and D are almost identical.

As measures of similarity, correlation coefficients fail to take account of the possibility of different elevations in the profiles and,

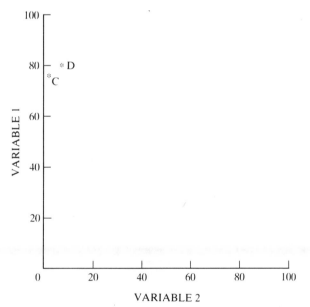

Figure 4.1   *Positions of models C and D plotted as points on statements 1 and 2.*

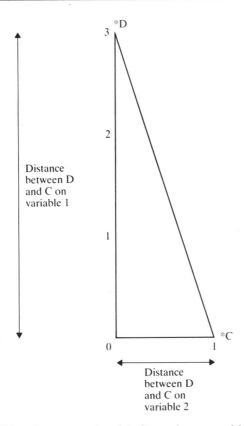

Figure 4.2 *Triangular representation of the distance between models C and D on statements 1 and 2.*

for this reason, other measures of similarity, particularly *distance*, tend to be preferred. Perhaps the most common measure of distance is known as *Euclidean* distance, which is the straight line distance between two objects when they are represented as points in hyperspace. It is a straightforward computational matter to calculate Euclidean distance. Conceptually, this is done as follows:

First of all, we plot the positions of models C and D on the 15 variables taken in pairs. Taking all pairs, we would make 15 × 14/2 plots = 105 plots! One such plot is shown in Fig. 4.1.

Figure 4.1 shows the positions of models C and D plotted as points on variables 1 and 2. On these two variables, C and D are manifestly very close to each other. We can see that on the axis representing variable 1, model D is displaced from model C by three percentage points, and on the other axis, representing variable 2, models C and D are displaced by one percentage point. So the distances between models C and D can be represented by the right angle triangle shown in Fig. 4.2.

Pythagoras informed us that 'the square on the hypotenuse of a right angle triangle is equal to the sum of the squares on the other two sides'. Thus, the square of the straight line distance between models C and D on variables 1 and 2 is equal to $3^2 + 1^2 = 10$. The Euclidean distance between models C and D is therefore equal to $\sqrt{10}$. To compute the Euclidean distance between models C and D across all 15 variables we compute the sums of squares of the differences for all variables taken in pairs, add them all together and then take the square root of the resultant number. Sometimes Euclidean distance measures are calculated as *average distance* since the distance between two objects usually increases as the number of variables on which they are compared increases. Also, when objects are compared on variables measured in different units of measurement, the distances between the objects are usually only calculated on the variables once they have been converted to standard scores. This ensures that all the variables are in the same units of measurement. If this is not done, then a variable (or variables) measured in 'small units' would have a disproportionate effect on the distance measure in comparison with variables measured in 'large units'.

Clearly, any two objects would be identical if the distance between them is equal to zero. So, in comparison with the correlation coefficient, the distance between objects is, in fact, a measure of their dissimilarity.

Euclidean distance is not the only possible measure of the distance between objects. For example, instead of applying Pythagoras' Theorem to arrive at a measure of the Euclidean distance between models C and D on variables 1 and 2, we could simply take the absolute difference between C and D, which is $3 + 1 = 4$. This measure of distance is known as *Manhattan* distance or *city-block*. Another popular measure of distance is known as *Mahalanobis*, D^2, also known as *generalized distance*. The definition of this distance measure needn't concern us here, but it is worth mentioning that unlike Euclidean or city-block distance, Mahalanobis incorporates the correlations between objects in arriving at a measure a distance. In this sense, Mahalanobis distance might be considered a more 'powerful' measure of distance in certain circumstances.

Principal coordinates analysis

No introduction to mapping techniques would be quite right without reference to geographical maps. A geographical map is a commonly used way of representing the distances between towns and cities. As well as indicating the distances between cities 'as the crow flies', with each city or town represented by a point, geographical maps also indicate the various routes from one city to another by representing the roads as lines. To help us to calculate the distance in road miles between towns and cities, organizations like the Automobile Association provide information about the distance between major towns and cities in the form of 'inter-city distance' tables. These tables are usually compiled by measuring the shortest road distance (or motorway distance) between the cities (so the distances are certainly not Euclidean). Now, what is of interest in the present context, is that by applying a mathematical technique known as *principal coordinates analysis* to a table of inter-city distances, it is possible to reproduce a map showing the relative positions of all the cities to each other. Figure 4.3 shows the map produced by applying principal coordinates analysis to a table of distances between forty-eight British cities provided by the Automobile Association in their handbook.

At first glance, the map in Fig. 4.3 appears to be a good approximation to a map of Great Britain—the outline is fairly familiar and the cities are in more or less the positions you'd expect them to be. However, the towns in the south-west are incorrectly located to each other in comparison with their positions on a typical geographical map of Britain. This is almost certainly due to the fact that the inter-city distance table from which the map in Fig. 4.3 was produced was a table of road distances and not a table of straight-line distances between the cities. Nevertheless, the map is a very reasonable representation and this is demonstrated by the fact that the two-dimensional configuration in Fig. 4.3 accounts for approximately 95 per cent of the variability in the original table of data. Figure 4.3 is in fact a remarkably good representation if we think about just how difficult it would be to 'draw' the map from the data in the original inter-city distance table, assuming, of course, we had

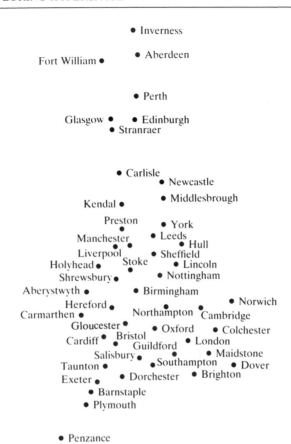

Figure 4.3   *Two-dimensional solution produced by applying principal coordinate analysis to a table of distances between 48 British towns.*

no a priori knowledge of the geography of Great Britain! Mathematically, as opposed to geographically, to explain the 'distances' between 28 cities requires a hyperspace of 27 dimensions. What principal coordinates analysis has been able to do, as it were, is to reduce 27 dimensions to just two and in the process it has lost only 5 per cent of the information in the original inter-city distance table. At this point, one further observation about Fig. 4.3 might be useful. Figure 4.3 has been presented in the familiar orientation of a map of Great Britain, that is, on axes pointing north/south and east/west. In actual fact, the points came off the computer with Aberdeen in a south-west position and Penzance in a north-east position. It was only our prior knowledge of the geography of Great

Britain that allowed us to rotate the map until it was in the familiar orientation. Moreover, it was our prior knowledge that allowed us to name the two dimensions as north/south and east/west. This issue—that of naming the axes or dimensions—will be picked up in our next example of principal coordinates analysis.

Before turning to a more typical use of principal coordinates analysis in survey research, to get a feel of how it works, it is most helpful to think of principal coordinates analysis as a variant of principal components analysis, which was discussed in the previous chapter. During the discussion of principal components analysis, it was suggested that the first principal component is the reference vector that accounts for the maximum amount of the common variability among a set of variables when they themselves are represented as vectors. In this sense, the first principal component is the 'line of best fit' to all the other variables represented as vectors.

Once the line of best fit has been drawn the original variables can then be defined in terms of their positions on this line. This can be done by projecting the ends of the (variable) vectors on the first principal component or line of best fit. Thus the positions of all the variables can be specified as points on the first principal component. In any instance, this form of representation will improve as the amount of variability accounted for by the first principal component increases. In our example in the last chapter, the amount of variability accounted for by the first principal component was 29.6 per cent. So, a representation of all 15 variables in one dimension as their projected points on one line (the first principal component) would not be a very good representation of the re-lationships in the original correlation matrix. An improved rep-resentation of the 15 variables would be provided by the projection of the variables onto the plane defined by the first two principal components, which would be represented by two vectors (axes) at right angles to each other. If we did this with the example used in the previous chapter, we would be able to represent (29.6 + 17.5) per cent of the variability in two dimensions, that is, 47.1 per cent. When principal components are represented graphically in this way, the original variables are represented as points in two dimensions and the distance between any two points is an approximation to the Euclidean distance between the 'points' at the ends of the original vectors (which represented the variables). In circumstances where the 'distance' between the original variables is not Euclidean, or where it cannot be reasonably assumed to be so, principal components analysis would not be an appropriate form of analysis.

Principal coordinates analysis, on the other hand, is not restricted to Euclidean distances and, moreover, works directly on distance or similarity matrices. It would seem, therefore, to have considerable advantages over principal components analysis when a visual representation of the data is desired. Certainly, in the example shown in Fig. 4.3, the original inter-city distances used were most definitely not Euclidean.

A fairly typical use of principal coordinates analysis in survey research is its application to the kind of data matrix exemplified by Table 2.4 in Chapter 2. This table shows the percentage of respondents who endorsed each of the 15 statements on the eight personal alarms plus the Ideal. Firstly, the similarities between the eight alarms and the Ideal are computed in terms of their distances from each other. In this example, the measure of inter-model distance computed was Euclidean, even though a generalized measure of distance (Mahalanobis) might be preferred. (The reasons for using Euclidean distance in this example will be made clearer when we discuss a technique known as the *biplot* later in this chapter.) Principal coordinates analysis was applied to the derived distance

Figure 4.4 *Two-dimensional solution produced by applying principal coordinate analysis to Euclidean distances between eight models on 15 statements.*

matrix to produce the two-dimensional solution shown in Fig. 4.4. This solution accounted for over 80 per cent of the variability, indicating quite a good fit.

Figure 4.4 is 'read' in a similar manner to the way we read a geographical map. The closer together two models appear on the map the more similar they are to each other in terms of their profiles on the 15 statements. Thus interpretation of the similarities between the models is a fairly straightforward matter. Models C and D are the two models that are most similar to each other in terms of their profiles across the 15 statements. Models A, B, H and the Ideal form another group or cluster and models E, F and G form another cluster. Model H was one of the new models and since it comes closest to the Ideal, it is the one new model that might be worthy of further development.

However, this map gives no obvious indication of why, specifically, the models have appeared in the positions where they are found on the map, so, unlike the map of Great Britain produced by principal coordinates analysis, we have no obvious way of naming the axes. However, by applying our 'knowledge' about the models —for example, the results of the factor analysis we carried out in the last chapter—we might begin to get some idea of the two major dimensions along which these products have been placed. Firstly, we know that models E, F and G are the larger models and that models C and D are smaller, less obtrusive alarms. If we 'project' the positions of the models on the horizontal axis (that is, drop perpendiculars from the points which represent the models), they fall out in the following order, taken from left to right:

E, G, F, H, Ideal, A, B, D, C

While this is not quite the same order we found when we profiled each of the models on factor 1 in Chapter 3, the same general pattern is found. We might therefore speculate that the horizontal axis is 'measuring' a dimension of size which taken from left to right is a dimension of large to small. Interestingly, the Ideal alarm is right in the middle, so surprise, surprise, the Ideal alarm should be neither too large nor too small.

Similarly, if we project the positions of the models on the vertical axis, from top to bottom, the order is:

D, C, E, F, G, A, H, B

From what we know already, we might speculate that this axis is measuring a dimension of 'hand-feel', ranging from 'poor hand-feel' at the top to 'good hand-feel' at the bottom.

Of course, it has to be said that naming the axes in this way can be largely a subjective exercise, even though, in the present case, we could argue that we have some objective or independent evidence. The question arises, therefore, as whether there are any more objective ways of explaining why the models fall out in their positions on the map. At least two such techniques are now available—the *biplot* and *correspondence analysis*. Both these methods produce two maps, one showing the similarities between the columns in a table (the alarms in our example) and the second showing the similarities between the rows in a table (the characteristics or statements in our example). The interesting point about the maps produced by these methods is that the positions of the models on the first map produced can be interpreted in the light of the positions of the characteristics on the second map produced. In other words, the biplot and correspondence analysis not only show the positions of the models but they also suggest which statements describe the models and 'cause' them to be placed in the positions found on the first map. Both methods set out to provide an objective explanation of why the models are similar or dissimilar to each other. How they do this will now be discussed.

The biplot

In Chapter 2, when we examined the data in Table 2.4, we noticed what were described as 'row effects' and 'column effects'; and to give us a better idea of which statements applied particularly to which models, we carried out what is called a two-way analysis of Table 2.4, to remove these effects. The results of this two-way analysis are shown in Table 2.6, the body of which is a table of 'residuals', that is, what is left, so to speak, when the row effects and the column effects have been removed. The residuals show, therefore, which statements apply to which models relative to the general or 'average' level of endorsement of all statements on all models. It was argued that an examination of the residuals provided a much clearer picture of the similarities between the models than an examination of the 'absolute scores' shown in Table 2.4.

In its most useful form in survey research, the biplot operates on a table of residuals akin to the ones shown in Table 2.6. To reinforce the point of why residuals are typically more informative than absolute scores, it is worth thinking about the kinds of data collected in market research studies when respondents are asked to rate objects such as brands on a number of statements. What is generally found in these circumstances is that well-known brands tend to receive a higher level of endorsement on most statements than brands that are less well-known. Indeed, it is sometimes argued that the level of endorsement of brands is directly related to their share of the market. In these circumstances, respondents' opinions of lesser known brands can be swamped by the well-known brands. By removing the row and column effects the perceptions of less well-known brands are allowed to surface, or so the argument runs. The first map produced by the biplot is a principal coordinates analysis like the one shown in Fig. 4.4. (In fact, the principal coordinates analysis shown in Fig. 4.4 was carried out on the distances calculated between residuals such as those in Table 2.6 and not on distances as calculated from the absolute scores, as was implied earlier.) This map has already been discussed so we can move on and consider the second map produced by the biplot.

This time, the biplot concentrates on the rows of residuals in Table 2.6. Firstly, it calculates the correlations between all the (variables) statements, taken in pairs, across the scores in the nine

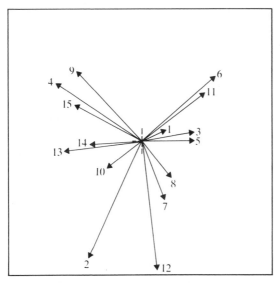

Figure 4.5   *Two-dimensional solution produced by applying principal components analysis to row residuals in Table 2.8.*

columns. (This is a reverse of the process used to produce the first map but in this instance we take the correlation coefficient between the statements as the measure of similarity rather than of the distance between them.) The resultant correlation matrix is then subjected to a principal components analysis and the results are plotted in the first two dimensions to give us the map shown in Fig. 4.5.

The first comment to be made about the representation in Fig. 4.5 is that the first two principal components account for over 80 per cent of the variability, providing us with a good fit of the original data. The variables are plotted as vectors in two dimensions and, as we know, the cosine of the angle between any two vectors represents the correlation between the two variables. Thus, for example, we can see that variables 6, 11, 1, 3 and 5 form a group of highly correlated variables. To refresh our memories, these are:

6. Could be easily kept in the pocket
11. Would fit easily into a handbag
1. Could be set off almost as a reflex action
3. Could be carried to be very handy when needed
5. Feels comfortable in the hand

These statements seem to be measuring the 'smallness' of the personal alarms. Interestingly, these five statement are most negatively correlated with statements 13, 14, 10 and 2 (remember the cosine of $180° = -1$) which when taken together seem to measure the 'largeness' of personal alarms, as follows:

14. Looks as if it would give off a loud noise
13. An attacker might be frightened that I might attack him with it
10. An attacker might have second thoughts about attacking me if he saw me with it
2. Would be difficult for an attacker to switch off

If we examine the remaining variables, we see that statements 9, 4 and 15 are highly correlated. Again, these statements relate to 'largeness' but also to the 'robustness' of the models. They are:

9. Solidly built
4. Would be difficult to break
15. I would be embarrassed to carry it round with me.

(Not surprisingly, statements 9, 4, 15, 14, 13, 10 and 2 tend to be more positively correlated with each other than with statements 6,

11, 1, 3 and 5, which, in turn, also tend to be more positively correlated with each other.)

Statements 8, 7 and 12 are also reasonably highly correlated with each other and they seem to refer to the 'ease of carrying', as follows:

8. I could keep a very firm grip of it if attacked
7. Would be difficult for an attacker to take it off me
12. Could be easily worn on the person

Rather than examining the intercorrelations between the statements in any further detail, the biplot is more useful when both maps are considered together.

Because the rows and columns in Table 2.6 are symmetric (i.e. their effects are additive) and because the maps produced in Figs 4.4 and 4.5 represent Euclidean distance between models and statements, respectively, the biplot allows us to interpret the results from the two maps directly. By 'scaling' the axes of the two maps so that they are equivalent, Figs 4.4 and 4.5 can be overlaid to produce the joint map shown in Fig. 4.6, allowing us to see directly which statements apply to which models.

To discover the relationship between a model and a statement,

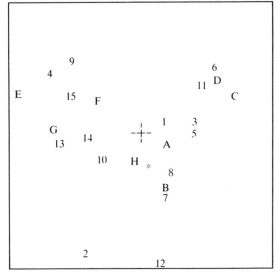

Figure 4.6 *Biplot of models and statements.*

we compute the product of three measures. The first measure is, literally, the straight-line distance between a model, represented in Fig. 4.6 by an alphabetic character (or, in the case of the Ideal, the symbol *), and the origin. The second measure is the straight-line distance between a statement, represented by a numeral in Fig. 4.6, and the origin. The third measure is the cosine of the angle between a model and statement as we've just described them. Given these definitions, the relationship between any model and any statement can be calculated by simple arithmetic, that is, by multiplying the three measures together. In more general terms, this means that the further a 'statement' is from the centre, the more it discriminates between the models. Similarly, the further a 'model' is from the centre, the more it is discriminated by any of the statements; and the smaller the angle between a 'statement' and a 'model', the more the 'statement' applies to the 'model'. 'Statements' that are opposite to any model apply to such a model in a negative sense (i.e. the cosine of 180° is equal to −1). In other words, we might be tempted to say, the converse of such a statement applies to such a model. Statements at right angles (orthogonal) to any model do not apply to any such model (since, as we know, the cosine of 90° is equal to zero). Of course, this does not mean that no respondents avowed that any statement described a model at right angles to it on the plot but, rather, that such a statement does not discriminate a model at right angles to it from any other model.

Given these general 'rules', we can also see that statements near the centre of the map do not disciminate between the models as well as statements further from the centre. For example, statement 1—'Could be set off almost as a reflex action'—is the least discriminating statement. By the same token, models nearer the centre of the plot are less discriminated by all the statements than models further from the centre.

An interpretation of Fig. 4.6 in the light of the foregoing considerations might be along the following lines:

> Existing models A and B and the new model H come closest to the Ideal. This cluster is defined in particular by statements 2, 7, 8 and 12. In other words, these were models which could be easily held in the hand and operated and which attackers would find difficult to switch off or to prise away from the victim.

The remaining three new models, E, F and G, form another cluster and are defined by statements 4, 9, 13 and 15. These models were the larger, robust models which could be used against the attacker.

The third cluster comprises existing models C and D and is defined by statements 6 and 11. These models were the smaller, less obtrusive alarms.

Finally, we should note that the statements which characterize the cluster containing models E, G and F, also characterize the cluster containing models A, B, H and the Ideal in a *negative* sense, i.e. the *converse* of statements 4, 9, 13 and 15, tend to apply to the Ideal personal alarm.

It is clear that for the personal alarm data, the biplot provides a very useful and informative visual display of the characteristics *and, at the same time*, of the models.

Correspondence analysis

A closely related technique to the biplot known as *correspondence analysis* or *reciprocal averaging* is another mapping technique which attempts to show the similarities between the rows and columns of a table of data like those collected in the personal alarms study. However, unlike the biplot, correspondence analysis does not allow one to make a direct and precise calculation of the relationship between rows and columns as in our case between statements and models. Instead, the two maps produced by correspondence analysis can only be interpreted together in a general directional sense. What is meant by this will become clearer as we apply correspondence analysis to the personal alarm data.

The version of the biplot described in the previous section was achieved by applying the biplot to a table of residuals (Table 2.6) produced by a two-way analysis of Table 2.4. Correspondence analysis works on a table of residuals produced by a chi-squared analysis, that is, it works on a table of residuals such as the one shown in Table 2.7. Like the biplot, correspondence analysis produces two maps. The first one, shown in Fig. 4.7, maps the similarities between the personal alarm models produced by a principal coordinates analysis of the distance between the models calculated from Table 2.7. The configuration of the models in Fig. 4.7 is broadly similar to the configuration produced by the biplot and shown in Fig. 4.4. Thus, we have the same three clusters of models, namely C and D; A, B, H and the Ideal; and E, F and G. Like the map produced by the biplot, the closer any two models are to each other on the map produced by correspondence analysis, the

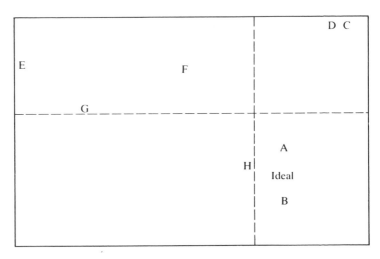

Figure 4.7   *Two-dimensional solution produced by applying principal coordinates analysis to Euclidean distance between eight models (from Table 2.6).*

more similar they are to each other in terms of their profiles on the 15 statements.

Again, the second map produced by correspondence analysis (analogous to the second map produced by the biplot) is a principal components analysis of the rows in Table 2.7. The principal components map of Table 2.7 produced by correspondence analysis is shown in Fig. 4.8. (This time the vectors have been omitted and the statements are shown as points at the ends of the vectors.)

Again, the correlations between the statements are represented by the cosines of the angles between vectors; and an inspection of the map in Fig. 4.8 indicates a very similar picture to the one shown in Fig. 4.5. Thus far, correspondence analysis has given rise to broadly similar results to those produced by the biplot. (Moreover, the maps produced account for over 80 per cent of the variance, just like those produced by the biplot.)

Like the biplot, the two maps produced by correspondence analysis can be overlaid to produce the map shown in Fig. 4.9. (Because the rows and columns are not symmetric, the origins of the maps are not in the centre as they are in the biplot.) Unlike Fig. 4.6, however, in which the positions of the models and statements can be interpreted precisely as the product of three measures, the positions of the models and the statements in Fig. 4.9 cannot be interpreted precisely. In Fig. 4.9, the relationships between models and state-

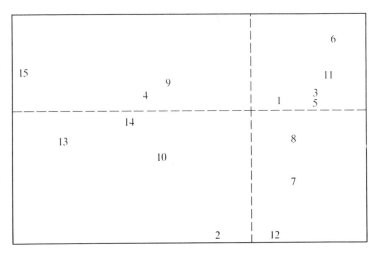

Figure 4.8 *Two-dimensional solution produced by applying principal components analysis to row residuals in Table 2.7.*

ments must be viewed in a general sense of direction. Thus, a model and a statement in a similar orientation to the origin are more closely related to each other than a model and a statement in different orientations to the origin. For example, in Fig. 4.9, we can say that statements 4, 9, 10, 13, 14 and 15 apply more to models E, F and G than they do to models C and D *because* models E, F and G and statements 4, 9, 10, 13, 14 and 15 are in the same general orientation to the origin, whereas models C and D and statements 4, 9, 10, 13, 14 and 15 are in a different general orientation to the origin. However, we cannot say from Fig. 4.9 that statement 15 'applies' more to model E than to statement 4, since statements 4 and 15 are both in the same orientation as model E to the origin. (If it were important to know which of statements 4 and 15 applied more to model E, this could, of course, be discovered from the biplot.)

In the personal alarms study, the Ideal alarm is conceptually the overall evaluative profile of the 15 statements against which the models were judged by the sample of respondents. In other words, in arriving at this profile, all respondents were treated as homogeneous in their evaluations of the statements. Potentially, the 'assumption of homogeneity' among respondents is a weak one since there could clearly exist distinct subgroups of respondents who evaluated the statements differently. Correspondence analysis can

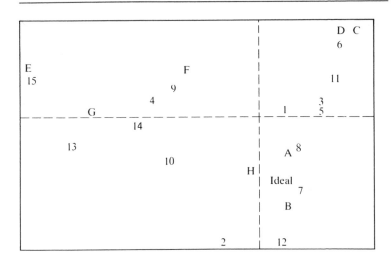

Figure 4.9   *Correspondence map of models and statements.*

be used to determine whether or not respondents do in fact form a homogeneous group in their evaluations of the characteristics. As an example, let's take a case where they do not. There will be a number of groups of respondents, X, Y, Z, for example, and a number of groups of statements, A, B and C. Respondents in group X may think that all the statements in A are good things for a personal alarm to possess and that none in group B is a good thing, whereas respondents in group Y might take the opposite view, and so on. Correspondence analysis can be used to order respondents and statements along a gradient so that respondents at the start of the gradient consider that statements at the start of the gradient are good things for a personal alarm to possess and later statements are not, whereas respondents at the end of the gradient do not think that statements at the start are good things, but that statements at the end are. Reordering the data matrix so that both respondents and statements are in the order of their gradient position then makes any discontinuities immediately apparent (see the example below). If the respondents do not fall into groups, there may well be a gradient of opinion with the highest and lowest scoring respondents disagreeing on all the statements, but with no clear discontinuities. A division into subgroups will, in this case, be arbitrary. Take, for example, the following data set:

Respondents	Statements							
	1	2	3	4	5	6	7	8
1	1	0	1	0	0	0	1	1
2	0	0	0	0	1	0	1	1
3	1	0	0	0	0	1	0	0
4	1	0	0	1	0	0	0	0
5	1	0	0	1	0	1	0	0
6	1	0	0	1	0	1	0	0
7	0	0	1	0	0	0	1	1
8	0	0	0	0	1	0	1	0
9	1	1	0	1	0	1	0	0
10	0	0	1	0	1	0	1	1

Little can be deduced from these data as they stand. However, if they are arranged with the respondents in the order 9, 5, 6, 3, 4, 1, 7, 2, 10, 8, and the statements in the order 2, 6, 4, 1, 3, 7, 8, 5, the existence of two groups of respondents becomes apparent.

Respondents	Statements							
	2	6	4	1	3	7	8	5
9	1	1	1	1	0	0	0	0
5	0	1	1	1	0	0	0	0
6	0	1	1	1	0	0	0	0
3	0	1	0	1	0	0	0	0
4	0	0	1	1	0	0	0	0
1	0	0	0	1	1	1	1	0
7	0	0	0	0	1	1	1	0
2	0	0	0	0	0	1	1	1
10	0	0	0	0	1	1	1	1
8	0	0	0	0	0	1	0	1

In this form, correspondence analysis simply projects respondents on the first axis, and any discontinuities would be shown by marked displacements of groups of respondents on the first axis (or gradient). In the personal alarms study, this form of correspondence analysis revealed that there were no marked discontinuities between respondents and the sample was accordingly regarded as homogeneous in its evaluations of the statements.

Of course, there is no reason why the biplot could not be used with the aim of achieving the same result. However, because of the nature of the distances calculated on the chi-square measure, any discontinuities there might be are more sharply identified by correspondence analysis. For example, when we project the models onto the first axis (the horizontal axis) in Figs 4.4 and 4.7, we can see that in Fig. 4.7, the models—E, G, F, H, Ideal, A, B, D and C—are more distant from each other than in Fig. 4.4. In other words, while both the biplot and correspondence analysis identify the same order of the models on the first axis, in correspondence analysis they are more sharply defined.

The biplot versus correspondence analysis

An obvious question that arises is, under what circumstances might the biplot be preferred to correspondence analysis and vice versa? The answer is that in most applications in survey research, the two approaches are largely interchangeable and they produce broadly comparable results. If the object in mapping the rows and columns in a table is primarily to explore the structure of the relationships, then it is difficult to see why one method might be *definitely* preferred over the other. However, from what's been said already, it is also true that—inherently—the biplot does have one advantage over correspondence analysis, namely, that it is possible to interpret the two maps together, precisely, whereas in correspondence analysis, as we saw, the relationship between the two maps is only a general directional one. For this reason alone, the biplot might be preferred. Moreover, for the same reason, if the objective in using either of these methods is to ask 'what if' questions of the sort 'what would happen to the positions of the models if statement *a* was thought to apply 20 per cent more to model *x* than it is perceived to currently?', it can be seen that the biplot, because of this inherent precision, would give a more sensitive answer. For this reason, one might be inclined towards the biplot. It has to be said, however, that correspondence analysis has a much wider application in survey research, owing simply to the fact that it has been promoted far more vigorously.

Non-metric multidimensional scaling (MDS)

Table 2.1 shows the average judged similarities between the eight personal alarms calculated from the individual judgements of each respondent. If you remember, respondents were presented with the models in pairs, such that every paired combination was covered, and they were asked to judge the similarity between the models by giving a score between '0' (no similarity) and '10' (identical). Strictly speaking, these sorts of judgements do not give rise to data that have absolute numerical significance. In other words, respondents' judgements should only be taken as *ordinal* judgements of the similarity between the models. The reason for this is that people in general can only make judgements of the kind, say, that Tom is taller than Dick, who is shorter than Harry and not that Tom is 1.26 times taller than Dick, who is 0.98 times shorter than Harry. Other than by chance, if people do try to make judgements of the latter sort, they are invariably wrong. Therefore, it is better not to treat the similarity scores in Table 2.1 as having a strict numerical significance but, rather as having only an ordinal significance. A variety of different techniques have been developed to map 'judged similarity' data and these are known generally as *non-metric multi-dimensional scaling methods* or MDS for short. They are known as 'non-metric' because they deal only with ordinal data. Principal coordinates analysis, on the other hand, is sometimes known as a *metric* or *classical* multidimensional scaling method because it deals with strict numerical data. Unlike these so-called classical methods, MDS provides a map of the similarities between objects on the basis of the rank ordering of the similarities alone. For example, Table 4.2 shows rank ordering of the 28 similarity measures between each pair of the eight models reported in Table 2.1.

Models C and D are ranked 1 (most similar to each other) and models C and E are ranked 28 (least similar to each other). The method works by first of all producing a configuration in which the models are arranged randomly. The rank order of the distances between these models configured at random is then compared with the rank ordering of the similarity data shown in Table 4.2. Initially, the 'goodness of fit' between these two rank

Table 4.2
Rank order of similarities between models (from Table 2.1)

	A	B	C	D	E	F	G	H
A	X							
B	4.5	X						
C	10	13.5	X					
D	8.5	17	1	X				
E	23	24	28	27	X			
F	12	13.5	22	20.5	8.5	X		
G	17	17	26	25	2	6	X	
H	3	4.5	20.5	15	19	7	11	X

orderings will usually be poor and the configuration is repeatedly moved around by a process of trial and error to achieve the best 'fit' possible. The goodness of fit is *estimated* by a measure known as *stress* which shows the relationship between the two rank orderings. Accordingly, after the first random configuration the *stress* will be very high but as the configuration is moved around to reduce the discrepancies the value of stress typically falls. The desired solution is obtained when stress is at a minimum. All MDS programmes work by minimizing the stress. If the rank order of the distances between the models in a configuration correlates perfectly with the rank orderings found in Table 4.2, the stress would be zero. Of course, in practice with the kind of data collected in the personal alarm study, we would not expect the stress to equal zero. What we want to achieve is as low a value of stress as possible. An informal evaluation of stress in terms of how well the resultant map fits the original similarity data is given below:

Stress (%)	Goodness of fit
20	Poor
10	Fair
5	Good
2½	Excellent
0	Perfect

One way of reducing the stress would be achieved simply by using more than two dimensions. (In our example of the similarities between eight personal alarms, a perfect representation, i.e. minimum stress, would be obtained in $8 - 1 = 7$ dimensions.) However, since the purpose of mapping is to represent the original hyperspace in two dimensions, this is not much use. The object, therefore, usually is to produce a two-dimensional configuration with minimum stress. Since MDS programmes produce the initial configuration at random, it is usually a sensible practice to 'run the data' several times using different randomly generated starting configurations on each occasion and to 'choose' the solution with the minimum stress.

Figure 4.10 shows the map produced from Table 2.1 with the minimum value of stress obtained from several runs of the data. The value of stress in this configuration is 'poor' according to our informal method of evaluation. Nevertheless, for completeness the map has been included. Though a 'poor' representation of the data, it still manages to reflect the relationships of similarity between the models that we've noticed before, particularly their projections on the first axis (the horizontal axis). Thus, E, G and F; H, B and A; and C and D tend to come together.

Even if the two-dimensional representation in Fig. 4.10 gave us a 'better' mathematical representation of the judged similarities shown in Table 2.1, we would still have the problem of trying to deduce why the models are configured in their relative positions to each other on the map. In other words, how do we describe or name the two axes? Frequently in studies where MDS techniques are used, the only data available concerning the similarities between the objects under study are judgements of overall similarity like the ones shown in Table 2.1. In these circumstances trying to work out

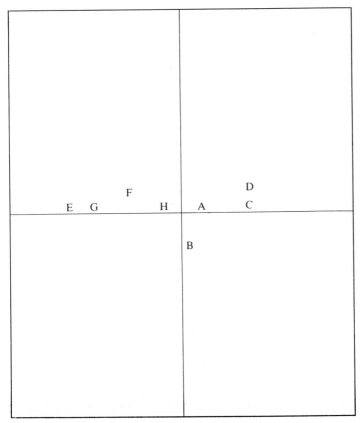

Figure 4.10   *Two-dimensional solution produced by applying MDS to the table of similarities shown in Table 2.1.*

what the two dimensions appear to represent can be a very speculative exercise. In our example in Fig. 4.10, given what we already know about the personal alarms, it is probably a safe guess to infer that the major dimension on which respondents were judging the similarities between the eight models was one of size. Certainly, the ordering of the models on the horizontal axis from left to right is such that the larger models are to the left of the origin and the smaller ones are to the right. This ordering is very similar to the orderings of the models on the first axes in the biplot and correspondence analysis. Moreover, there is so little displacement between the models on the vertical axis that we might be tempted to conclude that size was the only dimension on which the similarity between the models was being judged.

When objects have been rated on a set of attributes, as well as having been judged as to their overall similarity, it is possible to 'fit' these attributes on the MDS map as vectors, for example, such that the cosine of the angle between the attribute vector and the object vector multiplied by its length expresses the extent to which the attribute 'applies' to the object. When this is done, the attributes are usually referred to as *outside property vectors*. In our example, each of the eight models was evaluated on 15 statements, so in theory it would be possible to try to fit the 15 statements on the map in Fig. 4.10 as external property vectors. However, the map in Fig. 4.10 is such a 'poor' representation of the judged similarity between the models that doing so would be a rather pointless exercise. In any case, the same end result is achieved more effectively by the biplot and correspondence analysis. Nowadays, fitting attributes as vectors on MDS maps is rarely done. These techniques have been superseded by the biplot and correspondence analysis.

There is now a substantial amount of research evidence to show that different results from MDS are obtained depending on which features in the objects being judged as similar respondents are attending to. For example, in Fig. 4.10, we deduced that respondents were judging the overall similarity of the eight models in terms of their size. In other words, we inferred that the larger models were judged to be more similar to each other, overall, then they were to the smaller models. If respondents had been asked to judge the similarity of the models in terms of their 'hand-feel', we might have expected to obtain a different configuration from the one obtained in Fig. 4.10 by applying MDS to a table of 'judged similarities between the models on the basis of their hand-feel'. The point of this observation is to demonstrate that the concept of 'similarity' is an ambiguous one, psychologically. People do not judge the similarities between objects on the basis of an abstract notion of overall similarity. What they do is to judge the similarity between objects on the basis of certain features they share in common. The problem, therefore, in asking people to judge overall similarity is that there is no way of knowing on what characteristics the objects are being judged; and these may well vary from individual to individual. Indeed, in the example of MDS applied to Table 2.1, all the similarity judgements made by individuals were pooled and the analysis was carried out on the average data. As we have just indicated, there are good reasons to suppose the existence of individual differences, and unless a separate MDS analysis is carried out for each individual then such differences will be lost. This would not be satisfactory, though, since it would involve us in the difficult problem of assessing the degree of communality between one MDS solution and another. A better approach would be one in which individual judgements are pooled to provide an overall 'consensus' view, yet at the same time idiosyncracies of judgements are allowed to emerge.

One approach to assessing the degree of consensus within a sample of respondents has been to imagine something known as *normal attributes space* which conceptually contains *all* of the dimensions used by *all* of the respondents. Any one respondent could be using a subset of any or all of these dimensions when making judgements. For example, if two individuals are judging the similarity between pairs of bottles of wine, the judgement of one individual might be based on the dimensions of 'colour' and 'country of origin'. The judgement of the second might be based on 'price' and 'vintage'. The normal attribute space would therefore consist of four dimensions—colour, country of origin, price and vintage. It is now an easy matter to think of a third individual whose judgements might be based on, say, colour, country of origin and price. This individual would then share the same bases of judgement as the first individual and also show some overlap with the second. This conceptualization therefore allows for both communality in perception as well as the existence of purely idiosyncratic dimensions of judgements which may be used by a single individual. Of course, there is no reason to suppose that respondents would use each of the dimensions in an all-or-none manner. They may attach different degrees of importance to each of the dimensions.

The above reasoning has given rise to procedures that attempt to measure individual differences in respondents' perceptions of similarity. These procedures assume that individual differences can be interpreted in terms of respondents applying individual weights or saliences to the dimensions of a common space as described above. The object of these procedures is to recover the dimensions of this space—called the *group stimulus space*—and the weights attached to each dimension by each respondent. Accordingly, there are two kinds of output to be considered. The object space is the same as would be obtained from the MDS map produced from Table 2.1, for example. In addition to object space (e.g. the MDS map), these procedures produce a 'respondent space' in which respondents are represented as points in a space whose dimensions are the same as the object space. These reflect the differential importance of the dimensions to individual respondents, and the locations on these dimensions are the 'weights' or 'saliences' mentioned above. Of course, it's all very well to reflect the differences between individuals on the dimensions but the problem of identifying what these dimensions are still remains unanswered.

One final problem with MDS is worth mentioning before concluding this chapter on mapping. Mathematically, the concepts of 'similarity' and 'dissimilarity' are perfectly inversely related, yet quite different results have been obtained by applying MDS to data generated when respondents have been asked to judge the similarity between a set of objects and when they've been asked to judge the dissimilarity between the same set of objects.

Overall, therefore, the concept of similarity between objects is psychologically a tricky one which can lead to considerable difficulties in interpreting the results from MDS. Moreover, given the more recent advances in mapping represented by methods such as the biplot and correspondence analysis, MDS methods are probably mainly of interest on academic and historical grounds. Having damned MDS with some very faint praise, it is only fair to point out one area where its use is probably more helpful than either the biplot or correspondence analysis. This is in the area of sensory testing where, for example, the interest might be to ascertain the similarities between wines in terms of their taste, or perfumes in terms of their smell. In the latter example, respondents would be presented perfumes in pairs and asked to judge their similarities. The similarity data would then be subjected to MDS to produce a map of similar perfumes which could be related to preference measures for the perfumes under test. The reason why MDS is useful in sensory testing is probably that it is reasonably safe to assume that there are very few dimensions on which similarity is judged.

5. Cluster analysis

Background

At a superficial level, the basic idea upon which all methods of cluster analysis are based is very simple indeed. All of them attempt to group objects such as, in our example, models or respondents, into groups of similar objects, known as *clusters*. Objects are placed into different clusters such that members of any cluster are more similar to each other in some way than they are to members of any other cluster. Thus, methods of cluster analysis usually operate on measures of similarity or distance between objects; and these measures are typically defined and calculated in similar ways to those described in the previous chapter on mapping methods.

While the underlying idea of cluster analysis is appealing in its simplicity, there are a number of problems associated with it. As we noted in Chapter 2, when we attempted to compile Table 2.3 and draw the dendrogram shown in Fig. 2.2, the choice of which models should be placed into which clusters can be somewhat arbitrary since, for example, in these cases, the relative similarities between all the eight models were not clear-cut and we had to develop some arbitrary rules to group them together.

Indeed, different methods of cluster analysis—and there are now very many of them—can produce different solutions; that is, different clusters of objects from the same data set. The problem for the researcher is to discover which of the various solutions are the more indicative of any 'natural' groupings or clusters that may be in the data. To help in the task, a number of so-called validation techniques have been developed.

Perhaps the major problem with cluster analysis is the one also mentioned in Chapter 2, namely that cluster analysis *always* produces clusters of objects even in circumstances where there are no natural groupings in the data. In other words, although the aim of cluster analysis is to look for clusters of objects which reflect the structure in a set of data, it actually works by *imposing* a cluster structure on the data. This is all very well when natural clusters exist but it is likely to be misleading when there are no natural groupings. This is an extremely important point to grasp since cluster analysis applied to objects randomly distributed in hyperspace *will* produce clusters of them. Clearly, then, the success of using cluster analysis will depend entirely on knowing whether the clusters produced are *real* ones or simply artefacts of the method.

The foregoing comments beg a rather fundamental question to which, unfortunately, there is no satisfactory answer. This question is: 'what is a cluster?' Logically, if an entity cannot be defined adequately then it cannot be measured adequately, or alternatively if it had been measured adequately we would have no means of knowing that it had! Certainly, mathematicians are not very happy in circumstances where they cannot define their terms; and, in a strict sense, methods of cluster analysis are not really mathematical techniques at all. They belong to a group of methods known collectively as *heuristics*. Heuristics are sets of rules or algorithms which in the case of cluster analysis are drawn up and applied to data to create clusters of objects. By way of contrast, methods of factor analysis and mapping methods are based on mathematical procedures and in particular mathematical statistics, even though in the case of factor analysis, factors are typically 'chosen' on non-mathematical grounds. Nevertheless, many clustering routines do have certain mathematical properties which have been explored in detail in the statistical and mathematical literature. The important thing to note, however, is that cluster analysis methods essentially consist of sets of relatively simple rules which are intended to classify or assign objects into groups of similar objects.

The fact that the term 'cluster' cannot be defined precisely should not present too great a cause of concern. After all, there are many concepts in fields of enquiry such as biology, psychology and economics which have proved to be incredibly useful but which, like the term 'cluster', cannot be defined precisely. For example, very few biologists can agree entirely on what is meant by the term 'species', perhaps the most fundamental biological concept, yet this term has not inhibited the study of taxonomy and evolution; and only a complete fool would contend that nothing useful in biology has arisen simply because the term 'species' cannot be defined entirely adequately.

Even though a precise definition of what constitutes a cluster has proved to be elusive, most cluster analysts agree that clusters do have certain properties. Some of these will now be discussed.

Perhaps the most useful of these properties is that of *cluster variance*. If a number of objects form a cluster, then it is possible to calculate the centre of the cluster or the *centroid*, as it is sometimes called. The 'centre' of a collection of objects on one variable is the same as the average score of all the objects on the variable. With a normally distributed variable (see Chapter 1), typically, the mean would be taken as the centre; and the variance of the objects would be calculated as described in Chapter 1. On one variable, if we had two *distinct* clusters of objects, A and B, say, we would expect the mean score of objects in cluster A to be quite different from the mean score of objects in B. Moreover, we would not expect to find any overlap between the scatter of the scores around the mean of A and those around the mean of B. In other words, we would expect to discover a situation similar to the one shown in Fig. 5.1, if we had—truly—two *distinct* clusters, A and B.

From Fig. 5.1, it is fairly clear that the variance of the scores in cluster A and the variance of the scores in cluster B are much smaller than the variance of the scores overall, that is, the *total variance* of the scores when A and B are combined together as one group. Put another way, the variance *within* the clusters A and B is much smaller than the variance *between* clusters A and B, where the *between cluster variance* is defined as the *total variance* minus the *within cluster variance*. (NB   The total variance is calculated from

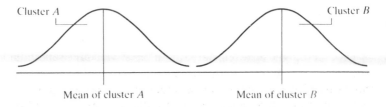

Figure 5.1   *Illustration of two distinct clusters on a single variable.*

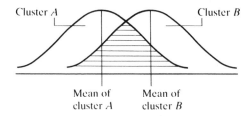

Figure 5.2   *Illustration of two overlapping clusters on a single variable.*

the deviations of all scores (i.e. *A* and *B* scores) about the *total mean* of all scores; and the within cluster variance is the sum of the variances of the scores within cluster *A* and those within cluster *B*, where the variance within cluster *A* is calculated from the deviations of the scores in *A* about the mean of *A* and the variance within cluster *B* is calculated from the deviations of the scores in *B* about the mean of *B*.)

Of course, it is comparatively unusual to find completely distinct clusters such as those depicted in Fig. 5.1. A more common situation is the one illustrated in Fig. 5.2, which shows two overlapping clusters on one variable, where the overlap is shown by the shaded area. The problem here is, to which of the two clusters should the objects falling in the shaded area be allocated? From Fig. 5.2, we can see that *all* the objects in the shaded area could be allocated to *either A* or *B*. One way of allocating the objects can be done on the basis of *maximizing the between cluster variance* while at the same time *minimizing the within cluster variance*. In the situation shown in Fig. 5.2 this could be achieved by allocating half the objects in the shaded area closest to the mean of cluster *A* to cluster *A* and the remaining half to cluster *B*. It can also be seen from Fig. 5.2 that the idea that two 'natural' clusters exist becomes more dubious the closer the means of *A* and *B* become, until in the limiting case the means of *A* and *B* coincide and we have one cluster only. (Even in the limiting case, however, a cluster analysis programme would still produce a two-cluster solution if we asked it to!)

The idea of the centroid of a cluster of objects derived on several variables can only be appreciated intuitively and not graphically, since, as we've noted in earlier chapters, once we have more than three variables, we cannot represent their interrelationships visually. For example, to represent the centroid of a cluster derived from six variables would require a hyperspace model in five dimensions in which the centroid was six points representing the mean scores of members of the cluster on the six variables in question.

In cases where some or all of the variables on which objects have been clustered are not normally distributed, the idea of variance as represented in our example does not hold true; and, typically in survey research, our centroids do not represent multivariate normal distributions. In these instances, cluster variance is probably better thought of as simply describing the relative nearness of points to one another in hyperspace. 'Good' clusters will be found when the within cluster variance is small, that is, when the points in the clusters are 'near' to each other; and 'bad' clusters will be found when the within cluster variance is large, that is, when the points in the clusters are 'far' from each other.

A notion closely related to that of cluster variance is the idea of *cluster radius*. If a cluster can be identified to have a round shape then it is known as a *hypersphere* and the radius of the hypersphere can be calculated as half the distance between the two members of the cluster furthest apart. Unfortunately, once identified, clusters do not always take on the shape of hyperspheres and so, generally, *cluster radius* is less useful than cluster variance. Indeed, many different kinds of cluster shapes have apparently been identified, so instead of thinking of the radius of a cluster, generally speaking, it's more useful to think of the *connectivity* of points in a cluster, which is the relative distance between the points in a cluster. This can also be estimated. Evidently, though, the idea of connectivity also breaks down with clusters that have particularly unusual shapes. All in all, the idea of cluster variance is probably the most useful one.

As mentioned at the beginning of the chapter, there are many different algorithms now available for performing cluster analysis. The most popular methods in survey research fall into two clusters of their own. One cluster is known by the collective name of *hierarchical agglomerative* methods and the other by the name of *iterative partitioning* methods. Both of these families of methods will now be considered; as usual, we'll use our data from the personal alarms study.

Hierarchical agglomerative methods

The dendrogram shown in Fig. 2.2 was produced by a hierarchical agglomerative method in which members were merged using a single-linkage rule (sometimes referred to as the *nearest neighbour* rule). All agglomerative methods sequentially merge, that is agglomerate, members according to a rule or set of rules (of which single-linkage is but one example) to produce a hierarchical organization that can be represented in a tree diagram or dendrogram such that at the lowest level the items being clustered are all independent and at the highest level are all joined into one group. If n items are to be clustered, all agglomerative methods require n − 1 steps to complete the clustering. Thus, seven steps were required to produce the dendrogram in Fig. 2.2 (Here $n = 8$, the number of models being clustered.)

All hierarchical agglomerative methods produce non-overlapping clusters that are nested, that is, each cluster is included or subsumed in larger clusters at higher levels of similarity. This is clearly illustrated in Fig. 2.2. One question that immediately arises, therefore, is whether it is reasonable to impose a system of nested clusters on the data in Table 2.1. Let's consider this question.

The dendrogram in Fig. 2.2 is only a *summary* of the data in Table 2.1. As we noted at the time of constructing Fig. 2.2 in Chapter 2, many of the original similarities between the models shown in Table 2.1 were lost when we applied the single-linkage rule. We need to know, therefore, how well the dendrogram in Fig. 2.2 represents the original similarities shown in Table 2.1. In other words we need some measure of the 'goodness of fit' of the dendrogram for the similarity matrix. 'Goodness of fit' can be judged by the extent to which the inter-model similarities shown in Table 2.1 are preserved by the dendrogram in Fig. 2.2. One way of estimating the similarities between any two items or objects as represented in a dendrogram is to take their similarities as the fusion level at which they both appear for the first time in the same cluster. For example, the 'implied' similarities between the eight models in Fig. 2.2 are those as shown in Table 5.1.

The goodness of fit of the dendrogram in Fig. 2.2 can now be estimated by comparing the original similarities in Table 2.1 with the 'implied' similarities shown in Table 5.1. One way of comparing the similarities in Tables 2.1 and 5.1 is to calculate the correlation coefficient between the two sets of data. In the jargon of cluster analysis, such a correlation coefficient is known as the *cophenetic correlation coefficient.* In our case the cophenetic correlation coefficient can be calculated as the product moment correlation, r. (To calculate r, the Xs in Table 2.1 are replaced by 10s to represent identity.) The value of r is equal to approximately 0.8. Generally

Table 5.1
Similarities between the eight models implied by the dendrogram in Fig. 2.2

	A	B	C	D	E	F	G	H
A	10.0	8.2	7.2	7.2	6.9	6.9	6.9	8.3
B	8.2	10.0	7.2	7.2	6.9	6.9	6.9	8.2
C	7.2	7.2	10.0	9.0	6.9	6.9	6.9	7.2
D	7.2	7.2	9.0	10.0	6.9	6.9	6.9	7.2
E	6.9	6.9	6.9	6.9	10.0	8.0	8.4	6.9
F	6.9	6.9	6.9	6.9	8.0	10.0	8.0	6.9
G	6.9	6.9	6.9	6.9	8.4	8.0	10.0	6.9
H	8.3	8.2	7.2	7.2	6.9	6.9	6.9	10.0

speaking, to be really confident that it is a valid procedure to impose a system of nested clusters on a set of similarity data, we would be looking for cophenetic correlation coefficients in excess of 0.9; and when the correlations fall below about 0.7, the assumption that the data consist of nested clusters becomes questionable. In our example, the evidence for the existence of nested clusters is not as convincing as we might wish but, nevertheless, we might be inclined to accept the assumption, given what we have already discovered about 'groups' of similar models from other analyses.

In circumstances where the imposition of a system of nested clusters is questionable, it is often useful to examine further the structure implied by the dendrogram. For example, the two-dimensional solution produced by applying MDS methods to the similarity matrix in Table 2.1 is shown in Fig. 4.10. Even though, as we noted, the stress value for this map is poor, by embedding the cluster solution of Fig. 2.2 on the map we can get a reasonably clear picture of the relationships between the eight models. This comparison is shown in Fig. 5.3, where we can see that models C and D form a 'tight' cluster, as do models H and A, and E and G. Moreover, the embedded cluster of H, A and B appears to be fairly distinct. The only model that seems to provide us with a problem is model F which is somewhat arbitrarily embedded in a cluster with models E and G.

Another check on the homogeneity of the clusters can be provided by comparing the mean similarity scores and standard deviations of members of each cluster with the mean of all the similarity scores for all models as indicated in Table 2.1. These figures are presented in Table 5.2. We can see that the members of each of the clusters have, on average, high inter-similarity scores and, moreover, there is little or no dispersion about the average, when compared with the overall (total) mean and its standard deviation. Furthermore, we can also see that the cluster containing C and D is the most tightly knit, followed by the cluster containing A, B and H. As we would expect, the third cluster, containing model F, is the least tightly knit of the three clusters. In terms of the idea of the 'between' and 'within' cluster variation mentioned earlier we can

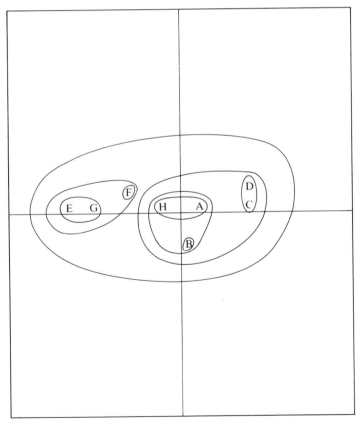

Figure 5.3 *Cluster solution implied by Fig. 2.2 embedded on the map shown in Fig. 4.10.*

Table 5.2
Means and standard deviations of similarity
scores of each cluster and overall

Clusters	Mean similarity	Standard deviation
C, D	9.0	0.0
A, B, H	8.23	0.12
E, F, G	7.87	1.22
Total	6.39	2.25

also see that the 'within' cluster variation is much smaller than the 'between' cluster variation.

As an alternative representation to the dendrogram in Fig. 2.2 clusters can be plotted in what are known as *icicle plots* since they are supposed to resemble icicles hanging from eaves. The icicle plot for the cluster solution as represented in Fig. 2.2 is shown in Fig. 5.4. The interpretation of Fig. 5.4 is along the same lines as the interpretation of the dendrogram in Fig. 2.2. At the lowest level—level 7—C and D are combined into a single cluster, giving seven clusters in all at this level. At the next level—level 6—six clusters are present and, so on, up to level 1, when all the models have been merged into a single cluster. So, for example, at level 3, there are three clusters and at level 2, there are two clusters. In certain ways, icicle plots are less informative than dendrograms; specifically, they give no information concerning the similarity level at which models are fused or merged.

It was mentioned earlier that the single-linkage rule for defining cluster membership is but one rule. There are many other commonly used linkage rules, each with their pros and cons, and it is certainly beyond the intended scope of this chapter to review the

Models

```
  F E G B A H C D
1 XXXXXXXXXXXXXXXXXXXXXXXX
2 XXXXXXX  XXXXXXXXXXXXX
3 XXXXXXX  XXXXXXX  XXXX
4 X  XXXX  XXXXXXX  XXXX
5 X  XXXX  X  XXXX  XXXX
6 X  XXXX  X  X  X  XXXX
7 X  X  X  X  X  X  XXXX
```

Figure 5.4 *Vertical icicle plot for Table 2.1 using single-linkage.*

relative merits of these different linkage rules. What is important to realize, however, is that different linkage rules can give rise to dendrograms of different shapes. Consequently, the application of different linkage rules to the same data set will sometimes suggest the existence of different clusters. Whenever possible, it is therefore prudent to examine further the structure of dendrograms produced by the application of a particular linkage rule. The cophenetic correlation coefficient is one way of looking at the structure, and comparing a dendrogram with a map produced by MDS on the same data is another way. In our example, both these examinations of the dendrogram in Fig. 2.2 would lead us to believe that the single-linkage rule has produced a reasonably good representation of the original similarities shown in Table 2.1.

One particularly useful application of hierarchical agglomerative methods is in the examination of the structure of correlation matrices such as the one shown in Table 3.2. In the previous chapter, it was suggested that the correlation coefficient between two variables can be considered as a measure of the similarity between the variables. Accordingly, it is reasonable to subject the data in Table 3.2 to a hierarchical agglomerative clustering method. An examination of the dendrogram or icicle plot so produced should give us a visual idea of the similarities (i.e. correlations) between the variables. In the present example, the data in Table 3.2 were clustered using a linkage rule known as *average-linkage* since experience has suggested that the average-linkage rule might give us a better representation than a dendrogram or icicle plot produced by a single-linkage rule or some other possible linkage rule. Average-linkage rules also vary but an example of one such rule is as follows. The algorithm begins by joining the two or more cases most similar to each other and then computes the average similarity score of this cluster. It then joins the next two cases most similar to each other, treating the first cluster formed as a single case with a similarity score represented by the average similarity score of its members. Thus, the second cluster formed could be an amalgam of the first cluster and one or more cases or it could be between two or more of the remaining cases after the first cluster has been formed. As clusters are merged their average scores are used to decide which cases should be joined together at successive levels.

The icicle plot produced from Table 3.2 using an average-linkage rule is shown in Fig. 5.5.

The cophenetic correlation coefficient takes a value of

Variables

```
   1 1 1 1 1     1
   3 5 2 1 4 9   0 4 8 7 6 3 2 5 1
 1 XXXXXXXXXXXXXXXXXXXXXXXXXXXXXXXXXXXXXXXXXXXX
 2 XXXXXXXXXXXXXXX     XXXXXXXXXXXXXXXXXXXXXXXXXX
 3 XXXXXXXXXXXXXXXX    XXXX  XXXXXXXXXXXXXXXXXXXX
 4 X   XXXXXXXXXXXXX   XXXX  XXXXXXXXXXXXXXXXXXXX
 5 X  X   XXXXXXXXXXX  XXXX  XXXXXXXXXXXXXXXXXXXX
 6 X  X   XXXX   XXXX  XXXX  XXXXXXXXXXXXXXXXXXXX
 7 X  X   XXXX   XXXX  XXXX  XXXX  XXXXXXXXXXXXXX
 8 X  X   XXXX   XXXX  X  X  XXXX  XXXXXXXXXXXXXX
 9 X  X   XXXX   XXXX  X  X  XXXX  X  XXXXXXXXXXX
10 X  X   XXXX   X  X  X  X  XXXX  X  XXXXXXXXXXX
11 X  X   XXXX   X  X  X  X  XXXX  X  XXXX   XXXX
12 X  X   XXXX   X  X  X  X  X  X  X  XXXX   XXXX
13 X  X   X  X   X  X  X  X  X  X  X  XXXX   XXXX
14 X  X   X  X   X  X  X  X  X  X  X  XXXX   X  X
```

Figure 5.5 *Vertical icicle plot for Table 3.2 using average-linkage.*

approximately 0.7 so, in line with the procedure mentioned earlier, MDS was applied to Table 5.2 and the cluster solution embedded on the map to give Fig. 5.6. Though the stress value for the map produced by MDS was poor, Fig. 5.6 gives a fairly clear idea of the relationships implied by the correlation matrix in Table 3.2. For example, Fig. 5.6 indicates that variables 10, 13 and 14 are unrelated to the remainder. These were the only three variables which did not load significantly on factor 1 in the unrotated principal components analysis described in Chapter 3.

Furthermore, variables 1 and 5, and 3 and 2 form fairly tight-knit clusters, and so do variables 7 and 8. The map also indicates that these three icicle plot clusters are not clearly separated from each other and, to a lesser extent, along with variable 6, all these variables tend to come together. This analysis sheds some light on the structure of factor 1 in the unrotated principal components analysis discussed in Chapter 3. Specifically, in descending order, variables 5, 1, 3, 2, 8, 6 and 7 are the highest loading variables on factor 1.

The icicle plot in Fig. 5.5 and the diagram in Fig. 5.6 can provide additional insights into the results obtained by factor analysis. The cynic might even say that factor analysis on its own provides a 'weaker' understanding of the correlations in Table 3.2.

Iterative partitioning methods

In commercial survey research, perhaps the greatest use of cluster analysis is in trying to identify clusters of respondents in terms of the

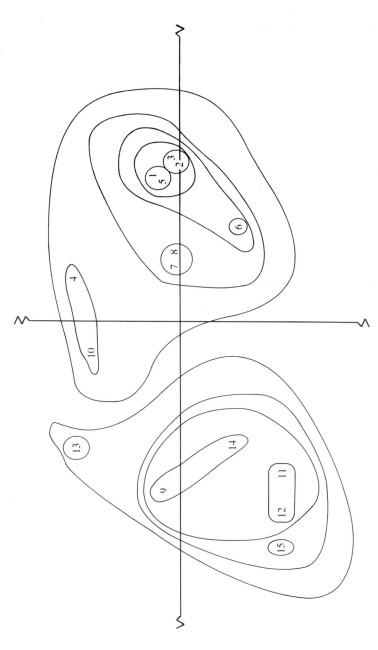

Figure 5.6   *MDS map for the correlation matrix in Table 3.2 with embedded average-linkage clustering solution.*

characteristics they share in common. The characteristics can be behavioural, attitudinal or a mixture of both. For example, it has become increasingly popular to attempt to *segment* populations of individuals into groups or clusters of individuals in terms of their 'lifestyles'. All these so-called lifestyle groups have been identified by the application of cluster analysis; and, almost invariably, the type of cluster analysis used has been from the family of techniques known as *iterative partitioning methods*.

Hierarchical agglomerative methods are usually unsuitable for clustering respondents for a number of reasons. Typically in survey research, relatively large samples of respondents are used and applying hierarchical agglomerative methods would necessitate the computation and storage of very large similarity matrices. For example, a sample of 1000 respondents would generate a similarity matrix containing approximately half a million values. These values would need to be stored in a computer's memory and repeatedly searched in an attempt to merge the respondents into clusters. Such a process would tax even the most sophisticated machines. Moreover, the prospect of trying to represent the similarities between such large numbers of respondents in dendrograms or icicle plots would be daunting in the extreme. Even if such plots could be done satisfactorily, an inspection—by eye—with the aim of identifying natural clusters of respondents would be almost impossible. For these reasons alone, when large samples of respondents are clustered, iterative partitioning methods are usually employed.

In very simple terms, the way these methods work is as follows. The data to be clustered are initially partitioned into a number of groups specified by the user. The clustering algorithm then attempts to rearrange the initial partition into the same number of clusters. For example, if a three-cluster solution of 900 respondents on ten variables is required, an iterative partitioning algorithm might begin by partitioning the data into three equal sized groups (300 in each group) at random. It would then compute the centroids of these three initial random groups on the ten variables. Next, the distances between each of the 900 respondents and the centroids of the three groups would be calculated. On the basis of these distances, respondents would be reassigned to the group with the nearest centroid. Upon completion of all the reassignments, the centroids of the new groups would be computed and the distances of all respondents from the new centroids calculated. Accordingly, respondents would again be reassigned to one of the 'new' groups with the nearest (new) centroid. This general procedure would be repeated, i.e. iterated, as many times as necessary to reach some arbitrary pre-set criterion, whence no more iterations are initiated and the clustering algorithm is complete. An example of an arbitrary, pre-set criterion might be to complete the process after n iterations, when the difference between the within cluster variance after n iterations is not significantly different from the within cluster variance after $n - 1$ iterations.

The major attraction of iterative partitioning methods is that they work directly on the raw data, thereby obviating the need to calculate the similarities between each and every respondent at the outset before clustering can begin. In principle, therefore, they can handle much larger data sets (i.e. larger samples) than hierarchical agglomerative methods. This apparent attraction does, however, have one major drawback. When the user initially specifies the number of clusters he eventually wants to produce, the question arises as how best to perform the initial partition into groups to eventually produce the most efficient cluster solution. In the simple description given above, it was suggested that the initial partition might be into three groups of equal size selected at random. Such a partition might be extremely inefficient, and in practice it has been demonstrated that it can produce 'local' clusters. These are local in the sense of being dependent almost entirely on the initial centroids that were selected. This point is often emphasized by demonstrating that different cluster solutions are usually produced by different initial partitions of the data. In theory, the best way of proceeding would be to partition respondents initially into all possible combinations of the three groups. Using each of these partitions in turn, the clustering process would be completed and the 'best' solution then chosen. Unfortunately, this intellectually appealing solution is computationally out of the question. Just think about it. With 1000 respondents, the number of possible starting partitions into three groups, of all possible sizes from 1 to 998 is literally millions of millions. Since this approach is manifestly impossible, a number of workmanlike approaches to the initial partition have been suggested. Some of these approaches appear to produce more efficient solutions than others but, unfortunately, there is no theoretical way of knowing whether one or more different initial partitions would have produced better solutions than the one actually adopted in a specific instance.

In addition to the problem of deciding on the initial partition, there is also the issue of deciding on what basis respondents should be moved from one group to another during the iteration process since the method we described in our simple description is by no means the only way. Finally, there is also the problem of deciding upon a criterion of when to cease iteration and complete the clustering process. Again, a number of different criteria have been suggested.

All of these issues have given rise to a confusing medley of different algorithms, the details of which will not concern us, except to emphasize that with data that do not contain 'natural' clusters, different algorithms tend to produce different solutions. Where the data being clustered do contain 'natural' clusters, by and large the different iterative partitioning methods tend to produce broadly similar results. Since often in survey research, we have no a priori reasons for suspecting 'natural' clusters, it oftens pays *not* to accept the results from one method of cluster analysis at face value but, instead, to attempt some form of replication by another method or methods. We'll turn to this later.

Finally, before turning to an example of an iterative partitioning method, one other question ought to be raised. In hierarchical agglomerative methods, the researcher examines the dendogram or icicle plot produced to decide on the number of clusters in the data, if any. In sharp contrast, in iterative partitioning methods the user specifies *in advance* the number of clusters he wants. The obvious question, therefore, is how does he know what number of clusters to specify? In survey research, the researcher often has no prior basis whatsoever to help him to specify a certain number of clusters. Usually, what he does is to specify a range of cluster solutions, say between two and ten, and he selects the 'best' solution, based on some independent criterion. We'll give an example of how this is done in our example below.

In Chapter 4 it was suggested that the possibility exists that our sample of respondents was not homogeneous in its perceptions of the 'Ideal' alarm. To test this hypothesis, a correspondence analysis was carried out in which respondents were 'plotted' on the first axis. The results of this analysis are shown in Fig. 5.7. Respondents (referred to as 'subjects') are ordered on a gradient. (The gradient positions are indicated by the column headed 'subject score' in Fig. 5.7) Respondents towards the start of the gradient are more likely to think characteristics 13, 9, 14, 10 and 5 (referred to as 'attribute numbers' in Fig. 5.7) are 'good' things for a personal alarm to possess, compared with respondents at the end of the gradient. In addition, respondents at the start of the gradient are less likely to think that characteristics 11, 7, 12, 4, 6, 2, 8, 3 and 1 are 'good' things for a personal alarm to possess than those 'further down' the gradient. Moreover, towards the middle of the gradient, respondents are more likely to think that all the characteristics (except characteristic 15) are good things for a personal alarm to possess. In other words, an examination of the gradient indicates a clear gradient of opinion from beginning to end. However, there are no clear discontinuities along the gradient. If there were, these would be indicated by 'large jumps' in the respondents' scores on the gradient. On the basis of examining Fig. 5.7, it would be an extremely difficult exercise to group respondents into clusters. There are clear differences of opinion along the gradient but partitioning respondents into separate groups or clusters according to their gradient positions would be arbitrary. Just where would one cluster begin and another end? Correspondence analysis has suggested there are no 'natural' clusters in the data. What does an iterative partitioning method of cluster analysis tell us? Before turning to this analysis, the fact that only one respondent—number 7—thought that characteristic number 15 was a 'good' thing, begs an explanation. It is hard to imagine that any respondent would think 'I would be embarrassed to carry it round with me' was a good thing for a personal alarm. This immediately suggests an error in the data. However, in checking back to the questionnaire it transpired it was not an error in data entry, suggesting, therefore, an error in coding the respondent's answer on the part of the interviewer. This error, however, does not have any effect on the overall conclusion we're inclined to draw from the correspondence analysis.

In specifying the cluster analysis on the 'Ideal' data, nine cluster solutions were requested, varying from a two-cluster solution up to a ten-cluster solution. As an initial examination of the data, the within cluster variance and the between cluster variance for each solution are examined. These variances are shown in Table 5.3, where each is expressed as a percentage of the total variance; and both, when added together, sum to 100 per cent, by definition. We can see that even in the ten-cluster solution, and a relatively small sample of 100 respondents, the within cluster variance is greater than the between cluster variance, suggesting that the data do not contain any 'strong' or tight natural groups. Moreover, we can also see that

Attribute number

Subject number	Subject score	15	13	09	14	10	05	11	07	12	04	06	02	08	03	01
7	★★★★★	★	★	★	★	★	★	-	-	-	-	-	-	-	-	-
33	−6.0	-	★	★	★	★	-	-	★	-	-	★	-	-	-	-
9	−5.0	-	★	★	★	-	-	★	-	★	-	-	★	-	-	-
3	−4.0	-	-	★	★	-	-	★	-	-	-	★	-	-	-	-
65	−3.1	-	-	-	★	★	★	-	★	-	-	-	-	-	-	-
31	−3.0	-	★	★	-	★	★	-	★	★	-	-	-	-	-	★
97	−3.0	-	★	-	★	★	-	★	-	★	-	★	-	-	-	-
17	−2.4	-	-	★	★	-	★	-	★	-	★	-	-	★	-	-
48	−2.2	-	★	-	★	★	★	-	-	★	-	-	-	★	-	★
37	−1.5	-	-	★	★	★	★	★	★	-	-	★	★	-	-	-
14	−1.8	-	★	-	★	★	★	★	-	★	-	-	★	-	★	-
44	−1.7	-	★	★	★	★	-	★	★	-	★	-	★	-	★	★
83	−1.2	-	★	★	★	★	★	★	★	★	-	★	★	★	★	-
86	−1.2	-	★	★	★	★	★	★	★	★	-	★	★	★	-	★
92	−.9	-	★	★	-	★	★	-	-	-	★	★	★	-	★	★
75	−.9	-	-	★	★	★	★	★	-	★	★	★	-	-	-	★
98	−.7	-	-	★	★	★	★	-	★	★	-	★	★	-	-	★
62	−.7	-	★	★	★	★	★	★	★	★	-	★	★	★	★	★
63	−.7	-	★	★	★	★	★	★	★	★	-	★	★	★	★	★
20	−.6	-	★	-	★	★	★	★	★	★	★	★	-	-	-	★
29	−.5	-	-	★	-	★	★	★	-	★	★	★	-	-	-	-
36	−.4	-	★	-	-	-	★	-	-	-	-	★	★	★	-	-
55	−.3	-	★	★	★	★	★	★	★	★	★	★	★	★	★	★
58	−.3	-	★	★	★	★	★	★	★	★	★	★	★	★	★	★
61	−.3	-	★	★	★	★	★	★	★	★	★	★	★	★	★	★
70	−.3	-	★	★	★	★	★	★	★	★	★	★	★	★	★	★
71	−.3	-	★	★	★	★	★	★	★	★	★	★	★	★	★	★
72	−.3	-	★	★	★	★	★	★	★	★	★	★	★	★	★	★
74	−.3	-	★	★	★	★	★	★	★	★	★	★	★	★	★	★
77	−.3	-	★	★	★	★	★	★	★	★	★	★	★	★	★	★
78	−.3	-	★	★	★	★	★	★	★	★	★	★	★	★	★	★
81	−.3	-	★	★	★	★	★	★	★	★	★	★	★	★	★	★
89	−.3	-	★	★	★	★	★	★	★	★	★	★	★	★	★	★
93	−.3	-	★	★	★	★	★	★	★	★	★	★	★	★	★	★
12	−.2	-	-	-	★	★	★	-	-	★	★	★	-	-	-	-
41	−.1	-	★	★	★	★	-	★	★	★	★	★	★	★	★	★
90	−.1	-	★	-	★	★	★	-	★	★	-	★	★	★	-	★
2	−.1	-	★	-	-	-	★	★	-	★	-	★	-	★	-	-
48	.0	-	★	-	★	★	★	★	★	★	★	★	★	★	-	-
8	.0	-	-	-	★	-	★	★	-	-	-	★	-	-	-	★

Figure 5.7a, b, c   *Correspondence analysis plotting respondents' positions on the first axis.*

```
                      Attribute number

Subject   Subject    1 1 0 1 1 0 1 0 1 0 0 0 0 0 0
number    score       5 3 9 4 0 5 1 7 2 4 6 2 8 3 1

50         .0        – – – ★ ★ ★ – – – ★ ★ – – – ★
28         .2        – ★ – ★ ★ ★ ★ – ★ ★ ★ – ★ ★ ★
13         .4        – – – ★ ★ – – – ★ – ★ – ★ – –
10         .4        – – ★ ★ – ★ ★ ★ ★ – ★ ★ – ★ ★
30         .5        – ★ – ★ – – – ★ ★ – ★ ★ ★ – ★
25         .5        – – ★ ★ ★ ★ ★ ★ ★ ★ ★ ★ ★ – ★
66         .5        – ★ – – ★ – ★ ★ – ★ – – ★ – ★
64         .6        – – ★ ★ ★ ★ ★ ★ ★ – ★ ★ ★ ★ ★
99         .6        – – ★ ★ ★ ★ ★ ★ ★ – ★ ★ ★ ★ ★
40         .7        – – ★ – ★ ★ – ★ – – ★ ★ – ★ ★

56         .8        – – – ★ ★ ★ ★ ★ – – ★ – ★ – ★
42         .8        – – ★ – ★ ★ ★ ★ – – ★ ★ ★ ★ –
67         .8        – ★ – – ★ ★ – – ★ – ★ ★ ★ – ★
27         .8        – – ★ ★ ★ ★ – ★ ★ ★ ★ ★ ★ ★ ★
88         .8        – – ★ ★ ★ ★ – ★ ★ ★ ★ ★ ★ ★ ★
49         .9        – ★ – ★ ★ ★ ★ ★ ★ ★ ★ ★ ★ ★ ★
 6         .9        – – ★ ★ ★ ★ ★ ★ ★ ★ ★ ★ ★ ★ ★
57        1.1        – – ★ – ★ ★ ★ ★ ★ ★ ★ ★ ★ – –
54        1.2        – ★ – – ★ ★ ★ ★ ★ – ★ ★ ★ – ★
26        1.2        – – ★ – ★ ★ ★ ★ ★ – ★ ★ ★ – ★

35        1.2        – – ★ – ★ ★ ★ ★ ★ – ★ ★ ★ – ★
51        1.4        – – – ★ ★ – – ★ ★ ★ ★ ★ – – –
96        1.4        – – – ★ ★ ★ – – ★ – ★ ★ – ★ ★
11        1.4        – – – ★ ★ ★ – – – ★ ★ – ★ ★ ★
39        1.6        – – – ★ ★ ★ – ★ ★ ★ ★ – ★ – ★
68        1.6        – ★ – – ★ ★ – ★ ★ – ★ ★ ★ ★ ★
22        1.6        – – ★ – ★ ★ ★ ★ ★ – ★ ★ ★ ★ ★
38        1.7        – – – ★ ★ ★ – ★ ★ – ★ – ★ ★ ★
76        1.9        – ★ – – ★ ★ ★ ★ ★ ★ ★ ★ ★ ★ ★
80        1.9        – ★ – – ★ ★ ★ ★ ★ ★ ★ ★ ★ ★ ★

87        1.9        – ★ – – ★ ★ ★ ★ ★ ★ ★ ★ ★ ★ ★
85        1.9        – ★ – – ★ ★ – ★ ★ ★ ★ ★ ★ ★ ★
59        1.9        – – ★ – ★ ★ ★ ★ ★ ★ ★ ★ ★ ★ ★
60        1.9        – – ★ – ★ ★ ★ ★ ★ ★ ★ ★ ★ ★ ★
79        1.9        – – – ★ ★ ★ ★ ★ ★ ★ ★ ★ – ★ ★
47        1.9        – – – ★ – ★ ★ – ★ – ★ ★ – ★ ★
 4        1.9        – – – – ★ ★ – – ★ ★ ★ – – – –
73        2.2        – – – ★ ★ ★ ★ ★ ★ ★ ★ ★ ★ ★ ★
82        2.2        – – – ★ ★ ★ ★ ★ ★ ★ ★ ★ ★ ★ ★
95        2.3        – – – ★ ★ ★ – ★ ★ ★ ★ ★ ★ ★ ★
```

Fig. 5.7 *continued*

		Attribute number
Subject number	Subject score	1 1 0 1 1 0 1 0 1 0 0 0 0 0 0 5 3 9 4 0 5 1 7 2 4 6 2 8 3 1
23	2.8	– – – – ★ – – ★ – ★ – – ★ – –
91	2.9	– – – – ★ ★ – ★ ★ – ★ ★ – – ★
1	2.9	– – – – ★ ★ – – – – ★ ★ – ★ ★
21	2.9	– – – – ★ ★ ★ – ★ ★ ★ – ★ – ★
53	3.0	– – – – ★ – – – – – – – ★ ★ –
15	3.0	– – – – – ★ – ★ – – ★ – ★ – –
32	3.0	– – – – ★ ★ ★ – ★ – ★ ★ – ★ ★
34	3.0	– – – – ★ ★ ★ ★ ★ – ★ ★ ★ – ★
69	3.1	– – – – ★ ★ – ★ ★ ★ ★ ★ – – ★
100	3.2	– – – – ★ ★ – – ★ – ★ ★ – ★ ★
52	3.2	– – – – ★ – – ★ ★ – ★ ★ – – –
45	3.4	– – – – ★ ★ – ★ ★ ★ ★ ★ ★ – ★
94	3.4	– – – – ★ ★ – ★ ★ ★ ★ ★ ★ – ★
5	3.4	– – – – ★ ★ – – ★ ★ ★ ★ – ★ ★
84	3.6	– – – – ★ ★ – – ★ ★ ★ ★ ★ ★ ★
16	3.8	– – – – ★ – ★ – ★ – ★ ★ – ★ ★
24	3.9	– – – – – ★ – – ★ – – ★ – ★ ★
43	4.0	– – – – – ★ ★ ★ ★ – ★ ★ ★ ★ ★
18	5.7	– – – – – – – – – – – ★ – ★ – ★
19	5.7	– – – – – – – – – – – ★ – ★ – ★

Fig. 5.7 *continued*

there is a fairly steady decrease in the within cluster variance as the number of cluster solutions increases. If there were any natural clusters, we might expect to see some marked decreases in the within cluster variance as the clustering algorithm discovered them. The *efficiency* of a cluster solution is sometimes defined as the between cluster variance expressed as a percentage of the total variance, since as the efficiency so-defined *increases*, the within cluster variance *decreases*. Obviously, by this definition, a high efficiency score implies tightly knit clusters with comparatively little variation around the cluster centroids. One way, therefore, of increasing the efficiency is to increase the number of clusters specified. In our example a 100 per cent efficiency would be achieved when all 100 respondents formed single clusters of their own, but a relatively large number of clusters defeats the very object of cluster analysis. Instead, what one usually hopes for is a reasonable level of efficiency, say above 40 per cent, with a manageable number of clusters (say, no more than ten) such that as the number

Table 5.3
Between and within cluster variance for two- to ten-cluster solutions
produced by an iterative partitioning method

Cluster size	Between cluster variance (%)	Within cluster variance (%)
10	46.1	53.9
9	42.1	57.9
8	42.8	57.2
7	40.2	59.8
6	36.1	63.9
5	31.9	68.1
4	28.3	71.7
3	20.3	79.7
2	18.8	81.2

of clusters increases beyond this number, only marginal increases in efficiency are obtained.

Even though our examination of the efficiencies of the various cluster solutions tends to confirm what the correspondence analysis has already suggested, for completeness let's examine one of the cluster solutions as an example. Purely for the purposes of exposition, we'll look at the two-cluster solution which, as indicated in Table 5.3, has an efficiency of 18.8 per cent. Cluster 1 contains 54 respondents and, accordingly, cluster 2 contains 46. One typical way of comparing clusters is to examine their profiles across the variables on which they have been clustered. In our example, this could be achieved by comparing the percentages of respondents in the two clusters who thought each of the 15 characteristics was a 'good' thing for a personal alarm to possess. In many cluster analysis applications, clusterings are carried out on variables measured on different scales. In these circumstances, the usual thing to do is to present cluster profiles on standardized scores. Cluster profiles across each variable are thus shown as deviations from zero means. In Fig. 5.8, the profiles of cluster 1 and cluster 2 are presented in this manner, where a single X represents 0.1 of a (unit) standard deviation from a zero mean. From Fig. 5.8 it can be seen that, on average, members of cluster 1 consistently score positively (i.e. above a zero mean) on all 15 characteristics except the last (problematical) one, whereas, on average, members of cluster 2 consistently score negatively (i.e. below a zero mean) on the first 14

Variables	Cluster 1			Cluster 2		
	−	0	+	−	0	+
1		. XXXX		XXXX .		
2		. XXXXX		XXXXXX .		
3		. XXXX		XXXXX .		
4		. XXX		XXXX .		
5		. XXXX		XXXXX .		
6		. XXXX		XXXX .		
7		. XXXXXX		XXXXXXX .		
8		. XXXXXX		XXXXXXX .		
9		. XXX		XXXX .		
10		. XXX		XXX .		
11		. XXXX		XXXXX .		
12		. XXXXX		XXXXX .		
13		. XX		XX .		
14		. XX		XX .		
15	X .				. X	

Key: X = 0.1 of unit standard deviation

Figure 5.8 *Cluster profiles for the two-cluster solution (×) = 0.1 of unit standard deviation.*

characteristics. The most marked differences are on characteristics 2, 7, 8 and 12, where the difference between the two clusters exceeds one standard deviation. In other words, while all the characteristics *tend* to discriminate between the two clusters, characteristics 2, 7, 8 and 12 are the *most* discriminating. Without our knowledge of the results of the correspondence analysis these two markedly different profiles might tempt us into believing we had identified two 'natural' groups of respondents. It is worth examining the results of the cluster analysis in the light of the correspondence analysis to see how these apparently contradictory results could have arisen.

In Fig. 5.9, Fig. 5.7 is reproduced with the members of cluster 1, as identified in the cluster analysis, in bold type so that we can see where they are located on the gradient produced by the correspondence analysis. An examination of Fig. 5.9 seems to indicate that the cluster analysis has tended to group into cluster 1 respondents who on average endorsed more characteristics overall, thus giving rise to the overall positive profile shown in Fig. 5.8. These tend to be the respondents towards the middle of the gradient. Respondents who endorsed fewer characteristics are mainly found towards the beginning and the end of the gradient and these respondents form the majority left in cluster 2, even though the patterns of endorsements of members of cluster 2 across the different characteristics vary markedly. Given this finding, perhaps we might expect the

```
                      Attribute number
                      1 1 0 1 1 0 1 0 1 0 0 0 0 0 0
Subject   Subject     5 3 9 4 0 5 1 7 2 4 6 2 8 3 1
number    score

  7       ★★★★★       ★ ★ ★ ★ ★ ★ - - - - - - - - -
 33        -6.0       - ★ ★ ★ ★ - - ★ - - ★ - - - -
  9        -5.0       - ★ ★ ★ - - ★ - ★ - - ★ - - -
  3        -4.8       - - ★ ★ - - ★ - - - ★ - - - -
 65        -3.1       - - - ★ ★ ★ - ★ - - - - - - -
 31        -3.0       - ★ ★ - ★ ★ - ★ ★ - - - - - ★
 97        -3.0       - ★ - ★ ★ - ★ - ★ - ★ - - - -
 17        -2.4       - - ★ ★ - ★ - ★ - ★ - - ★ - -
 48        -2.2       - ★ - ★ ★ ★ - - ★ - - - ★ - ★
 37        -1.5       - - ★ ★ ★ ★ ★ ★ - - ★ ★ - - -

 14        -1.8       - ★ - ★ ★ ★ ★ - ★ - - ★ - ★ -
 44        -1.7       - ★ ★ ★ ★ - ★ ★ - ★ - ★ - ★ ★
 83        -1.2       - ★ ★ ★ ★ ★ ★ ★ ★ - ★ ★ ★ ★ -
 86        -1.2       - ★ ★ ★ ★ ★ ★ ★ ★ - ★ ★ ★ - ★
 92         -.9       - ★ ★ - ★ ★ - - - ★ ★ ★ - ★ ★
 75         -.9       - - ★ ★ ★ ★ ★ - ★ ★ ★ - - - ★
 98         -.7       - - ★ ★ ★ ★ - ★ ★ - ★ ★ - - ★
 62         -.7       - ★ ★ ★ ★ ★ ★ ★ ★ - ★ ★ ★ ★ ★
 63         -.7       - ★ ★ ★ ★ ★ ★ ★ ★ - ★ ★ ★ ★ ★
 20         -.6       - ★ - ★ ★ ★ ★ ★ ★ ★ ★ - - - ★

 29         -.5       - - ★ - ★ ★ ★ - ★ ★ ★ - - - -
 36         -.4       - ★ - - - ★ - - - - ★ ★ ★ - -
 55         -.3       - ★ ★ ★ ★ ★ ★ ★ ★ ★ ★ ★ ★ ★ ★
 58         -.3       - ★ ★ ★ ★ ★ ★ ★ ★ ★ ★ ★ ★ ★ ★
 61         -.3       - ★ ★ ★ ★ ★ ★ ★ ★ ★ ★ ★ ★ ★ ★
 70         -.3       - ★ ★ ★ ★ ★ ★ ★ ★ ★ ★ ★ ★ ★ ★
 71         -.3       - ★ ★ ★ ★ ★ ★ ★ ★ ★ ★ ★ ★ ★ ★
 72         -.3       - ★ ★ ★ ★ ★ ★ ★ ★ ★ ★ ★ ★ ★ ★
 74         -.3       - ★ ★ ★ ★ ★ ★ ★ ★ ★ ★ ★ ★ ★ ★
 77         -.3       - ★ ★ ★ ★ ★ ★ ★ ★ ★ ★ ★ ★ ★ ★

 78         -.3       - ★ ★ ★ ★ ★ ★ ★ ★ ★ ★ ★ ★ ★ ★
 81         -.3       - ★ ★ ★ ★ ★ ★ ★ ★ ★ ★ ★ ★ ★ ★
 89         -.3       - ★ ★ ★ ★ ★ ★ ★ ★ ★ ★ ★ ★ ★ ★
 93         -.3       - ★ ★ ★ ★ ★ ★ ★ ★ ★ ★ ★ ★ ★ ★
 12         -.2       - - - ★ ★ ★ - - ★ ★ ★ - - - -
 41         -.1       - ★ ★ ★ ★ - ★ ★ ★ ★ ★ ★ ★ ★ ★
 90         -.1       - ★ - ★ ★ ★ - ★ ★ - ★ ★ ★ - ★
  2         -.1       - ★ - - - ★ ★ - ★ - ★ - ★ - -
 46          .0       - ★ - ★ ★ ★ ★ ★ ★ ★ ★ ★ ★ - -
  8          .0       - - - ★ - ★ ★ - - - ★ - - - ★
```

Figure 5.9a, b, c *Correspondence analysis shown in Fig. 5.7 with members of cluster 1 in bold type.*

Attribute number

Subject number	Subject score	15	13	09	14	10	05	11	07	12	04	16	02	08	03	01
50	.0	–	–	–	★	★	★	–	–	–	★	★	–	–	–	★
28	.2	–	★	–	★	★	★	★	–	★	★	★	–	★	★	★
13	.4	–	–	–	★	★	–	–	–	★	–	★	–	★	–	–
10	.4	–	–	★	★	–	★	★	★	★	–	★	★	–	★	★
30	.5	–	★	–	★	–	–	–	★	★	–	★	★	★	–	★
25	.5	–	–	★	★	★	★	★	★	★	★	★	★	★	–	★
66	.5	–	★	–	–	★	–	★	★	–	★	–	–	★	–	★
64	.6	–	–	★	★	★	★	★	★	★	–	★	★	★	★	★
99	.6	–	–	★	★	★	★	★	★	★	–	★	★	★	★	★
40	.7	–	–	★	–	★	★	–	★	–	–	★	★	–	★	★
56	.8	–	–	–	★	★	★	★	★	–	–	★	–	★	–	★
42	.8	–	–	★	–	★	★	★	★	–	–	★	★	★	★	–
67	.8	–	★	–	–	★	★	–	–	★	–	★	★	★	–	★
27	.8	–	–	★	★	★	★	–	★	★	★	★	★	★	★	★
88	.8	–	–	★	★	★	★	–	★	★	★	★	★	★	★	★
49	.9	–	★	–	★	★	★	★	★	★	★	★	★	★	★	★
6	.9	–	–	★	★	★	★	★	★	★	★	★	★	★	★	★
57	1.1	–	–	★	–	★	★	★	★	★	★	★	★	★	–	–
54	1.2	–	★	–	–	★	★	★	★	★	–	★	★	★	–	★
26	1.2	–	–	★	–	★	★	★	★	★	–	★	★	★	–	★
35	1.2	–	–	★	–	★	★	★	★	★	–	★	★	★	–	★
51	1.4	–	–	–	★	★	–	–	★	★	★	★	★	–	–	–
96	1.4	–	–	–	★	★	★	–	–	★	–	★	★	–	★	★
11	1.4	–	–	–	★	★	★	–	–	–	★	★	–	★	★	★
39	1.6	–	–	–	★	★	★	–	★	★	★	★	–	★	–	★
68	1.6	–	★	–	–	★	★	–	★	★	–	★	★	★	★	★
22	1.6	–	–	★	–	★	★	★	★	★	–	★	★	★	★	★
38	1.7	–	–	–	★	★	★	–	★	★	–	★	–	★	★	★
76	1.9	–	★	–	–	★	★	★	★	★	★	★	★	★	★	★
80	1.9	–	★	–	–	★	★	★	★	★	★	★	★	★	★	★
87	1.9	–	★	–	–	★	★	★	★	★	★	★	★	★	★	★
85	1.9	–	★	–	–	★	★	–	★	★	★	★	★	★	★	★
59	1.9	–	–	★	–	★	★	★	★	★	★	★	★	★	★	★
60	1.9	–	–	★	–	★	★	★	★	★	★	★	★	★	★	★
79	1.9	–	–	–	★	★	★	★	★	★	★	★	★	–	★	★
47	1.9	–	–	–	★	–	★	★	–	★	–	★	★	–	★	★
4	1.9	–	–	–	–	★	★	–	–	★	★	★	–	–	–	–
73	2.2	–	–	–	★	★	★	★	★	★	★	★	★	★	★	★
82	2.2	–	–	–	★	★	★	★	★	★	★	★	★	★	★	★
95	2.3	–	–	–	★	★	★	–	★	★	★	★	★	★	★	★

Fig. 5.9 *continued*

```
                      Attribute number
                      1 1 0 1 1 0 1 0 1 0 0 0 0 0 0
Subject    Subject    5 3 9 4 0 5 1 7 2 4 6 2 8 3 1
number     score

  23         2.8       - - - - ★ - - ★ - ★ - - ★ - -
  91         2.9       - - - - ★ ★ - ★ ★ - ★ ★ - - ★
   1         2.9       - - - - ★ ★ - - - - ★ ★ - ★ ★
  21         2.9       - - - - ★ ★ ★ - ★ ★ ★ - ★ - ★
  53         3.0       - - - - ★ - - - - - - - ★ ★ -
  15         3.0       - - - - - ★ - ★ - - ★ - ★ - -
  32         3.0       - - - - ★ ★ ★ - ★ - ★ ★ - ★ ★
  34         3.0       - - - - ★ ★ ★ ★ ★ - ★ ★ ★ - ★
  69         3.1       - - - - ★ ★ - ★ ★ ★ ★ ★ - - ★
 100         3.2       - - - - ★ ★ - - ★ - ★ ★ - ★ ★

  52         3.2       - - - ★ ★ - - ★ ★ - ★ ★ - - -
  45         3.4       - - - - ★ ★ - ★ ★ ★ ★ ★ ★ - ★
  94         3.4       - - - - ★ ★ - ★ ★ ★ ★ ★ ★ - ★
   5         3.4       - - - - ★ ★ - - ★ ★ ★ ★ - ★ ★
  84         3.6       - - - - ★ ★ - - ★ ★ ★ ★ ★ ★ ★
  16         3.8       - - - - ★ - ★ - ★ - ★ ★ - ★ ★
  24         3.9       - - - - - ★ - - ★ - - ★ - ★ ★
  43         4.0       - - - - - ★ ★ ★ ★ - ★ ★ ★ ★ ★
  18         5.7       - - - - - - - - - - ★ - ★ - ★
  19         5.7       - - - - - - - - - - ★ - ★ - ★
```

Fig. 5.9 *continued*

three-cluster solution to pull cluster 2 apart into two separate clusters. In fact, it does. The three-cluster solution leaves cluster 1 intact but divides cluster 2 into one cluster containing 45 respondents and another containing only one respondent—our friend, respondent number 7—who was the only respondent who reportedly endorsed characteristic number 15. The four-cluster solution then splits the cluster of 45 respondents into two further clusters, at the same time taking four respondents away from our original cluster 1, but still leaving respondent number 7 in a cluster on his own. However, in terms of their gradient positions, the allocation of respondents into different clusters is still rather arbitrary and the most sensible conclusion to draw is that the data do not contain any natural groups.

A very basic approach to validating the existence of natural clusters of respondents in survey research is to profile the members of each cluster on some other variables that were not included in the clustering process. For example, if a sample of respondents is clustered on its attitudes to an object and four clusters of respondents with different attitudes to the object in question are tentatively identified, then these clusters could be profiled on demographic characteristics such as age and gender. If the four clusters thus profiled revealed, say, that cluster 1 comprised mainly young men, cluster 2 comprised mainly young women, cluster 3 comprised mainly older men and cluster 4 comprised mainly older women, then we might be tempted to believe the four clusters had a degree of *external validity*. In practice, it is comparatively rare to find really clear-cut differences on external variables even though some researchers profile clusters on every possible external variable, thereby inevitably increasing the chances of finding some significant differences! Profiling clusters on external variables in a *post hoc* fashion can only be considered as a minimum step towards validity.

Choosing the variables on which to cluster

A question that is often raised in using cluster analysis is 'which variables should be chosen as the basis for clustering?' One approach favoured by some researchers is to use as many as possible in the hope that if enough are chosen the clustering process will discover some inherent structure in the data. Clearly the more variables that are input into a clustering process, the more likely the process is to discover some sort of structure, so most experts would not recommend such a 'naïve' empirical approach. It is much better to have a clear idea beforehand (or a theory, even) as to the types of variables that might distinguish between groups or clusters of individuals and use these rather than putting everything into the 'computer-pot', so to speak, and hoping for the best. Even when some considered thought has been given to the types of variables that might be entered into a cluster analysis, there is still a problem in dealing with large numbers of variables, since large numbers of variables can give rise to clusters that are difficult to interpret, especially if there is no 'theory' to guide the interpretation. It has therefore become fairly common practice in survey research to attempt a reduction of large numbers of variables to a smaller but representative subset before commencing cluster analysis. Factor analysis is frequently used when variables are known to be inter-correlated. The reasoning behind this approach to reducing the number of variables is fairly sound in the following sense. If highly correlated variables are used in a cluster analysis, then in computing a measure of similarity using these variables the effect is to give an implicit weight to the variables. For example, if two variables are very highly correlated, then the effect of using both of them in a cluster analysis is to give twice the weight to them compared with other variables that are not correlated.

Principal components analysis can be used to reduce the 'weighting' or dimensionality in the data, producing derived variables or factors to use as input to a cluster analysis. Unfortunately, there is some disagreement among the experts about the wisdom of adopting this procedure but it is fairly common practice in survey research. Large batteries of attitudes' statements are usually factor analysed to produce a reduced number of factors. When respondents are clustered on these factors they are assigned *factor scores* to replace the large number of scores on each individual attitude statement. An individual respondent's factor scores are composite scores derived from summing the products of the factor loadings of each of the attitude statements with the respondent's score on each statement.

Concluding remarks

One of the major thrusts of this chapter has been concerned with the validity of cluster solutions, that is, whether the clusters 'identified' by the various algorithms represent natural groups of objects or whether clusters are merely *produced* (rather than identified) as a consequence of the clustering rules used. This is perhaps the major issue in cluster analysis.

Unfortunately, the example of iterative partitioning clustering given in the chapter did not produce a 'natural grouping', so the question left in the air is, 'what does a cluster solution that has identified natural clusters look like?' There is no straightforward answer to this question. If natural clusters of respondents do exist, one thing we might expect to notice is a marked decrease in the within cluster variance once the algorithm had identified them. This point was mentioned in discussing Table 5.3 a few pages back but it is probably as well to re-emphasize it.

Another approach to identifying natural clusters with large samples of respondents has been to look at the *reliability* of the cluster solution as opposed to the validity. Provided the sample of respondents is large enough (say 1000 or more), it can be divided at random into two samples each with equal numbers of respondents. Cluster solutions can then be computed for each split-half sample.

Providing the cluster program is not 'bound' to give similar results (this is dependent on the choice of the initial centres), then if the two split-half samples produce very similar solutions, we might be inclined to think we have discovered something real. Unfortunately, reliability is not the same thing as validity (for example it is possible to have a valid but unreliable measuring instrument, such as a broken thermometer purporting to measure temperature, and a reliable but invalid measuring instrument, for instance, a working thermometer purporting to measure barometric pressure). In our example where we tried to cluster respondents on the Ideal, if we had had a large enough sample with the same pattern of responses as we found in our small sample, it is conceivable that a split-half run would have produced two cluster solutions similar to the one we obtained. This would not mean, however, that we had found valid, natural clusters but simply that the clusters we had found could be reproduced, i.e. that they are 'reliable'.

Recipients of cluster solutions should always be very wary about the clusters presented to them and survey analysts should take great care before claiming they have discovered any natural clusters.

Index

Abstract expressionism, 13
Acute angle, 21
Angle between two Vectors (*see* Correlation)
Arithmetic average (*see* Mean)
Automobile Association Handbook, 29, 87
Average distance, 86
Average-linkage rule, 119

Between cluster variance, 112–113
Biplot, comparison with correspondence analysis, 102
  explanation and example of, 92–97
Bipolar factors, 65
British Crime Survey, 27

Cage, John, 13
Causation (*see* Correlation)
Centroid (*see* Cluster)
Census, 10
Chi-squared analysis, 45
City-block distance (*see* Manhattan distance)
Classic multidimensional scaling (*see* Principal coordinates analysis)
Clusters:
  centroid of, 112
  connectivity of points in, 114
  hypersphere as shape of, 114
  idea of, 111–112
  identification of, 112–13
  profiles of members of, 129–130
  radius of, 114
  variance, 112–113
Cluster analysis:
  compared with correspondence
analysis, 124–134
  compared with non-metric multidimensional scaling, 116, 120
  hierarchical agglomerative methods of, 114–120
  idea of, 110
  iterative partitioning methods of, 120–129
  of a correlation matrix, 119–120
  problems with, 110–112, 135
  use of factor scores in, 135
Cluster solutions:
  efficiency of, 128
  external validity of, 134
  reliability of, 136
  validation of, 116–118, 133–134
Column effects, 38–39
Common factor variance, 51, 61–63
Common value, 39
Communality, 52, 63–64
Confidence limits, 10–13
Connectivity (*see* Cluster)
Cophenetic correlation coefficient (*see* Goodness of fit)
Correlation:
  and causation, 22–23, 79
  and factor loadings, 50, 59
  as cosine of angle between vectors, 20–21
  as measure of similarity, 82–83
  concept of, 15
  geometrical representation of, 20–22
  magnitude of, 16
  matrix, 53, 55
  significance of, 54
Correspondence analysis:
  comparison with cluster

Correspondence analysis – *cont.*
  analysis, 124–134
  comparison with biplot, 102
  explanation and example of,
    97–102
Cosine of angle (*see* Correlation)

D² (*see* Mahalanobis distance)
Dendrogram, 32–34
Deviation, from mean, 3
Dispersion (*see* Scatter)
Disraeli, Benjamin, 24
Dissimilarity (*see* Similarity)
Distance, measures of, 85–87

Efficiency (*see* Cluster solution)
Eigenvalue (*see* Extracted
    variance)
Einstein, Albert, 9
Euclidean distance, 85–86
Evolutionary theory, 13
External validity (*see* Cluster
    solution)

Factor:
  extraction criteria, 64–65
  interpretation of, 65–69
  loadings, 50, 59
  represented as reference
    vectors, 50, 59–61
  rotation, 69–75
  scores, 135
  score weights, 75–78
  significance of, 66
Factor analysis:
  derived solutions, 70
  description of, 50–65
  direct methods of, 69
  limitations of, 78–80
  qua principal components
    analysis, 48
  typical application in survey
    research, 49
Finnegans Wake, 13
Four-fold point correlation, 53

Gambler's fallacy, 14
Generalized distance (*see*
    Mahalanobis distance)

Goodness of fit:
  cophenetic correlation
    coefficient as measure of, 115
  stress, as measure of, 104
Group stimulus space, 108

Heuristics, 111
Hierarchical agglomerative
    methods (*see* Cluster
    analysis)
Histogram, 1
Hypersphere (*see* Clusters)
Hyperspace, definition of, 57

Icicle plots, 118
Interpreting factor loadings,
    65–69
Iterative partitioning methods (*see*
    Cluster analysis)

Joyce, James, 13

Latent root (*see* Extracted
    variance)
Leading diagonal, of correlation
    matrix, 54
Loading (*see* Factor loading)

Mahalanobis distance, 86–87
Manhattan distance, 86
Mapping methods, ideas behind, 82
Margin of error (*see* Confidence
    limits)
MDS (*see* Non-metric
    multidimensional scaling)
Mean, 1
Median, 7–8
Metric multidimensional scaling
    (*see* Principal coordinates
    analysis)
Mode, 7–8
Multidimensional space (*see*
    Hyperspace)

Nearest neighbour rule (*see*
    Single-linkage rule)
Non-metric multidimensional
    scaling, explanation of,
    102–109

compared with cluster analysis, 116, 120
Normal attribute space, 107
Normal distribution, 5

Oblique angle, 21
Oblique rotation of factors, 70
Obtuse angle, 21
Ordinal judgements, 103
Ordination methods (*see* Mapping methods)
Orthogonal angle, 21
Orthogonal factors, 61, 69
Orthogonal rotation of factors (*see also* Varimax rotation), 70
Outside property vectors, 106

Personal alarms, concept of 'Ideal', 35
description of project, 26–28
Plato's Academy, 24
Principal components analysis:
description of (*see* Factor analysis)
relation to principal coordinates analysis, 89–90
qua factor analysis, 48
Principal coordinates analysis:
examples of, 87–89, 90–91
explanation of, 89–90
relation to principal components analysis, 89–90
Pythagoras, 86

Quantum mechanics, 13

R^2, square of correlation coefficient, 20, 61–63
Radius (*see* Clusters)
Random, randomness and random sampling, 13–15
Range, 1
Reciprocal averaging (*see* Correspondence analysis)
Reference vector (*see* Factor)
Regression analysis, 15
Reliability (*see* Cluster solutions)
Residuals (*see* Row residuals)
Rotation (*see* Factor)

Row effects, 38, 39
Row residuals, 39

Samples and sampling, 8–15
Scatter, 2
Scatterplot, 15
Sensory testing, use of non-metric multidimensional scaling in, 108–109
Significance, of correlations, 54
of factor loadings, 66
Statistical (*see* Statistical significance)
Similarity:
and dissimilarity, 108
direct measures of, 83
idea of, 82–83, 107
indirect measures of, 83
matrices, 29, 30, 31, 32
Single-linkage rule, 33
Standard deviation, 4
Standard error, of mean, 9–10
Standard scores, 6, 7
Statistical significance, 12–13
Statistics, limitations of, 24–25
Stress (*see* Goodness of fit)
Sums of squares (*see* Extracted variance)
Table effects, 39
Table of cosines, 18–19
Table of residuals, 39
Thought experiment (*see* Einstein)
Tree diagram (*see* Dendrogram)
Two-way analysis, 39

Unipolar factors, 65
Upper quartile, 54
Use of table residuals in market research, 92–93

Validity (*see* Cluster solutions)
Variability, index of, 3
Variance, 3
Varimax rotation, 70

Within cluster variance, 112–113

Z scores (*see* Standard scores)